T0238290

Lecture Notes in Computer Science 771

Edited by G. Goos and J. Hartmanis

Advisory Board: W. Brauer   D. Gries   J. Stoer

Gerald Tomas   Christoph W. Ueberhuber

Visualization of
Scientific Parallel Programs

Springer-Verlag

Berlin Heidelberg New York
London Paris Tokyo
Hong Kong Barcelona
Budapest

Series Editors

Gerhard Goos
Universität Karlsruhe
Postfach 69 80
Vincenz-Priessnitz-Straße 1
D-76131 Karlsruhe, Germany

Juris Hartmanis
Cornell University
Department of Computer Science
4130 Upson Hall
Ithaca, NY 14853, USA

Authors

Gerald Tomas
Christoph W. Ueberhuber
Institute for Applied and Numerical Mathematics, Technical University Vienna
Wiedner Hauptstraße 8-10/115, A-1040 Wien, Austria

CR Subject Classification (1991): D.2.2, D.1.3, D.1.7, G.1.0, G.1.4, G.1.7, G.4

ISBN 3-540-57738-6 Springer-Verlag Berlin Heidelberg New York
ISBN 0-387-57738-6 Springer-Verlag New York Berlin Heidelberg

CIP data applied for

© Springer-Verlag Berlin Heidelberg 1994
Printed in Germany

Typesetting: Camera-ready by author
SPIN: 10131031   45/3140-543210 - Printed on acid-free paper

Preface

The substantial effort of parallel programming is only justified if the resulting codes are adequately efficient. In this sense, all types of performance tuning are extremely important to parallel software development. With parallel programs, performance improvements are much more difficult to achieve than with conventional (sequential) programs. One way to overcome this inherent difficulty is to bring in *graphical* tools.

When trying to visualize parallel programs, one is faced with a vast amount of relevant literature consisting of more than 100 articles in journals or conference proceedings and dozens of software systems.

This book pursues two major goals: first, to cover recent developments in parallel program visualization techniques and tools, most of which have not yet been dealt with by text books; second, to demonstrate the application of specific visualization techniques and software tools to scientific parallel programs.

For this purpose, two prototypical problem areas have been chosen: (i) *solution of initial value problems of ordinary differential equations* – a notoriously difficult problem with respect to parallelization, and (ii) *numerical integration* – a problem which is seemingly easy to parallelize. In these two fields the advantages of parallel program visualization are demonstrated. One representative software system – PARAGRAPH – is used to show how visualization techniques can contribute to experimental performance assessment and improvement.

Synopsis

Part I of this book is a general section which introduces the topic of parallel program visualization and gives an overview of current techniques and available software tools.

Part II describes a particular class of parallel methods for the numerical solution of initial value problems of ordinary differential equations.

Iterated Defect Correction (IDeC) is an acceleration technique which iteratively improves numerical approximations to the solution of differential equations. The initial numerical approximation is obtained by using a (one-step or multi-step) discretization method.

The family of IDeC methods allows various kinds of parallelization. The most important design decision has to be made with respect to load distribution. The calculations associated with the basic method and the solution of the neighboring problems can be allocated to the available processors in different ways. The investigation of the relative merits of these variants using the parallel program visualization system PARAGRAPH is described in Part II of this book.

Part III deals with parallel integration algorithms. These algorithms have attracted the interest of researchers ever since the early days of multiprocessor systems. This interest can be attributed to the beneficial property of additivity with respect to the partitioning of the region of integration into pairwise disjoint subregions. Nevertheless, scalable parallel integration algorithms with high efficiency are not easy to obtain. This is particularly true for problem adaptive algorithms.

The most troublesome feature of the otherwise attractive globally adaptive integration schemes is their inherently sequential nature. Major algorithmic modifications and extensions are necessary to make efficient implementations possible on parallel computers. Even more significant is the effort needed to obtain satisfactory performance over a wide range of target machines.

Parallel program visualization helps to find satisfactory load distribution schemes which are instrumental to the utilization of the available power of a parallel computer. In particular, dynamic load distribution mechanisms are often much easier to assess when their effect can be studied in pictorial representations.

Acknowledgements

We would like to thank our colleagues Roman Augustyn, Michael Karg and Arnold Krommer at the Technical University of Vienna for having produced some of the material included in Parts II and III.

We are also grateful to Michael Heath (University of Illinois) and Martin Schönhacker (Technical University of Vienna) for their valuable comments on the manuscript.

We wish to thank the members of the Institute for Statistics and Computer Science of the University of Vienna, especially Gabriele Kotsis, for providing many of the abstracts in the bibliography.

Finally, we wish to express our gratitude to the Austrian Science Foundation for continued financial support under Grant No. S 5304–PHY.

January 1994   GERALD TOMAS and CHRISTOPH UEBERHUBER

Contents

III Visualization of Parallel Integration Methods 221

13 Parallel Integration   225

14 Simulated Target Machines   229

15 Trace File   231

16 Integration-Specific Displays   235

17 Visualization of Parallel Integration Algorithms   237

Bibliography   261

Index   305

Part I

Parallel Program Visualization

Chapter 1

Introduction

Computer programs are often difficult to understand when viewed in their traditional textual form. Any textual representation may obscure the functionality of a program. Consequently, program visualization research has been motivated by the desire to explain the functioning of algorithms by means of animated displays.

Humans can process tremendous quantities of image information in parallel, detecting and tracking complex visual patterns with incredible speed (Arbib, Robinson [1]). Nevertheless, as the amount of presented information grows, the viewer's ability to understand the resulting image can become rapidly saturated unless the displayed information's level of abstraction is increased. For this reason abstraction plays an important role in any type of visualization. By providing flexible abstractions, a visualization system can help its user in selecting and specifying pictorial representations that can easily be understood.

Program visualization research was originally motivated by the desire to explain the processing of sequential algorithms by means of graphical displays. With the advent of modern computer architectures, it became evident that the difficulties in understanding programs are even more pronounced when the target machine is an advanced parallel computer. Thus, visualization can play a key role in the software development process for parallel and distributed computer systems.

An important motivation for writing and using parallel and distributed programs instead of sequential programs is to achieve better performance of some kind. Significant speed-ups and/or larger problem-solving abilities must be achieved in order to prove the cost-effectiveness of the additionally required software and hardware investments, not to mention the increased programming overhead. Unfortunately, the development and improvement of techniques and tools for the performance evaluation and the debugging of parallel programs lag behind the progress of modern hardware technology.

1.1   Debugging of Parallel Programs

Debugging parallel programs is in many ways different from debugging sequential ones. It is not only the added complexity which makes the debugging process more difficult: traditional sequential debugging is a *cyclic debugging*, i.e., a repeated stopping of the program at breakpoints, examination of the current program state, and then either continuation or re-execution in order to stop the execution at some other point. This conventional style of debugging is often not feasible with parallel programs.

The major difficulties of debugging parallel and distributed programs are (McDowell, Helmbold [13]):

Increased Complexity: The explicit expression of concurrency introduces new dimensions of complexity. Consequently, the development and debugging of parallel programs is more difficult than writing or debugging their sequential counterparts.

Nonrepeatability (Nondeterminism): Parallel programs do not always show reproducible behavior. Even when run with the same input data, their control flow and even their results may differ dramatically. Such differences are caused by *races*, which occur whenever two activities are allowed to progress in parallel. For example, one process may attempt to write data into a specific memory location while a second process is reading from that memory cell. The behavior of the second process may differ radically, depending on whether the old or the new value is read. Therefore, the cyclic debugging approach inevitably fails for such parallel programs because any such undesirable behavior may not appear again when the program is re-executed. Any type of undesirable behavior with a low probability of occurrence will hardly ever be detected in test runs.

Probe Effect: Any attempt to observe the behavior of a distributed system may change the behavior of that system. This phenomenon represents the *Heisenberg Uncertainty* applied to the field of software evaluation, otherwise known as the *probe effect*. For instance, any additional debugging statement may drastically change the behavior of programs with races. The probability of detecting interesting (i.e., undesirable) program behavior may be reduced seriously by the probe effect.

Any kind of nondeterminism is particularly difficult to deal with because the programmer often has little or no control over it. For instance, the result of a race may depend on the current load of the processors involved, the amount of network traffic, and specific nondeterminism of the communication channels.

Lack of a Synchronized Global Clock: Without a synchronized global clock it may be difficult to determine the precise order of events occurring in distinct, concurrently acting processors (Lamport [9], Stone [116]).

1.2 Approaches to Parallel Debugging

Possibilities for debugging parallel software are (McDowell, Helmbold [13]):

Extending Traditional Debugging: The simplest type of a parallel debugger is (or seems to behave like) a collection of sequential debuggers. The primary differences lie in the user dialog and multiple debugger instances.

All the difficulties already mentioned (probe effect, influence of nondeterminism, etc.) may cause serious problems. However, often those effects do not occur, because many parts of parallel programs are *not* time-dependent in their course of computation.

Another serious problem is that simple debuggers usually operate at a very low level. A major difficulty emerges with programs with many concurrently executed processes. The interprocess level of such programs is completely ignored by these types of debuggers.

Event Histories: During the execution of a distributed program, *events* are recorded in an *event history* (also called *trace file*), which can be examined after the program's termination. The definition of an event requires the time it occurred, its type, and some specific additional information (for instance, the processor number or the message length). Note, however, that an "event" is a flexible concept: an event may be a memory access, a message sent or read, an access to a data object, a task activity, or may even be some user-defined incident.

An event history is often a very large collection of data. Accordingly, most debuggers provide facilities to browse or query the event history in a systematic way.

Some systems, like BUGNET (Wittie [122]) or INSTANT REPLAY (LeBlanc, Mellor-Crummey [11]), have the capability to *replay* a parallel program according to an event history, i.e., they are able to re-execute the program with the same (relative) order of events. In this way they allow the reproduction of previous (possibly erroneous) results. If the event history contains all relevant information, a single process can be debugged in isolation from the remainder of the program. All necessary communication is provided by the event history.[1]

---

[1] A prerequisite for this type of debugging is an event history which contains all messages received from and sent to other processes as well as external input/output events and interrupts.

Use of Graphics: There are several fundamental techniques used by many parallel debugging systems for displaying information.

> **Textual Representation** of program code and data may involve color, highlighting, or a dynamic display of control flow information.

> **Time Process Diagrams (Gantt Charts)** present program data in a two-dimensional way with time on one axis and individual processors on the other.

> **Animation** of the program execution symbolically places each process (or portion of distributed data) on an individual point in a two-dimensional display. The display corresponds to a single instant of time (*snapshot*). As time advances, the display changes like an animated movie.

> **Multiple Windows** permit several simultaneous views of one parallel program, which allows the identification of different types of errors.

Static Analysis: Sometimes static analysis techniques are used for detecting certain classes of anomalies in parallel programs (for example, McDowell [12]). This approach is distinct from a formal proof of correctness because no attempt is made to demonstrate conformance with some kind of written specification. Instead, static analysis techniques try to insure that the investigated program cannot enter certain predefined states that generally indicate errors.

With static analysis mainly two kinds of errors can be detected:

> **Synchronization Errors,** which include deadlocks and wait-forever, and

> **Data-Usage Errors,** which occur when uninitialized variables are read and when two processes simultaneously update a shared variable.

1.3   Performance Tuning

In debugging parallel programs, detecting errors and correcting them is the most important step. A second step is required, however, to bring about a satisfactory performance. A debugged parallel program will eventually produce correct results, but its performance may be very poor. One reason may be an inefficient algorithm that wastes processor time or other resources. The process of detecting such algorithmic bottlenecks is called *performance evaluation*; the process of improving program performance by adapting algorithms or finding better, more efficient ones is called *performance tuning*.

The use of sequential tools for performance tuning is, again, not a good choice. Of course, sequential tools can be used to detect performance bottlenecks of an algorithm executed on a single processor, but these sorts of bottlenecks are not as serious as performance degradation caused by badly organized, or even erroneous, interprocess communication.

Event histories are valuable not only for debugging purposes, but also in the process of performance tuning. To what extent event histories can reveal internal performance problems depends on the kind of data they contain. The minimum information required is when each processor was working, when it was not working, and why it stopped or resumed its work.

The gathered information is presented preferably in graphical form. For instance, the *Space-Time Diagram* used in PARAGRAPH (Heath, Etheridge [57]) graphically indicates when each processor was working (solid line) or not working (blank), and where messages were sent (indicated by lines connecting the involved nodes). Multiple windows make it possible to visualize various performance aspects from different views.

1.4 Visualization of Parallel Programs

In both debugging and performance tuning of parallel programs, the information to be analyzed may be extremely voluminous (including states of processes and interactions between processes) and often describes interrelations of considerable complexity among processor activities, data objects and interprocess communication. Consequently, tools for trace management are required which are able to discard events which, for instance, took place outside a given time interval, and thus give their users an opportunity to filter out irrelevant events.

There is a distinction to be made between a *view* and a *visualization*: different *views* of an investigated program can be constructed by emphasizing or ignoring selected information. For each view, there is a great number of ways to represent the information to be analyzed. Any *visualization* makes some details of a view manifest while obscuring others. A view defines *what* information is presented, a visualization describes *how* the information is displayed (LeBlanc et al. [72]).

A single view can be sufficient neither for debugging nor for performance tuning. Consequently, every useful program analysis environment must allow the examination of different aspects of program behavior by providing different views. Any particular view may contain either too much or too little information to explain the relevant aspects of the program's behavior. Unless the environment's user knows exactly what to look for, interesting phenomena are likely to be lost in an enormous amount of confusing or badly arranged information.

Recent efforts in program analysis have often been problem-oriented, resulting in tools that address only a specific class of problems. Many of these tools are only useful for one particular programming paradigm or programming language, or they do not adapt to other types of errors or performance aspects. This lack of generality seriously limits their applicability. For example, Brewer et al. [32] developed a tool to detect memory access patterns of sequential Fortran programs, and COMET (Kumar [8]) measures the highest possible parallelism inherent in sequential Fortran programs. Both tools are limited to programs written in Fortran *without* any parallel extensions.

Program analysis would be much easier if a user-selected subset of the available tools could be assembled and used as one individually defined tool. However, as most tools are architecture or language specific, it is difficult or impossible to combine them with others. Even an unstructured collection of tools would not create a solution for this problem because the user would lack the guidance necessary to choose which tool to apply in a particular situation.

A very useful property of visualization tools would be their adaptiveness. *Adaptive tools* are structured in such a way that users can select views of their programs in some adequate succession, where each view takes into account previous views and the user's current needs.

Chapter 2

Visualization Tools

During the last few years visualization has attracted increasing attention in many fields. But there are great differences in what and how some kinds of information can be visualized. This chapter reviews different types of visualization found in currently available tools and systems.[1]

2.1 Types of Visualization

Visualization can be divided into three general fields (Myers [92], Roman, Cox [105]):

Scientific Visualization refers to the graphical representation of application oriented data such as that produced by supercomputer simulations, satellites, or measurement devices used in fields such as astronomy, meteorology, geology, and medicine.

Program Visualization comprises techniques where the program is specified in a conventional, textual manner, and pictorial representations are used to illustrate different aspects of the program, for instance, its run time behavior. Program visualization systems were initially used for debugging or teaching and presentation purposes. In the context of advanced computer systems, program visualization is additionally used for performance monitoring and tuning.

Visual Programming refers to any system that allows the user to specify a computer program in a two-(or higher-)dimensional, graphical form.

---

[1]The tools and systems mentioned in this chapter are meant to serve as examples and are not a complete list of tools of a particular category.

In this context conventional languages cannot be categorized as two-dimensional because the respective programs are treated as one-dimensional character streams by their corresponding compilers or interpreters. Following the same logic, systems based on conventional (linear) programming languages whose main function is the definition of pictures (like Postscript, HP-GL, etc.) or the generation of drawings (like Mac-Draw, etc.) are *not* to be subsumed under visual programming because they are not able to create computer programs.

Visual programming systems give non-programmers or unskilled programmers the possibility of creating fairly complex programs after only a short period of training. Visual programming is also a useful technique for producing user-friendly, easy to learn graphical user-interfaces (GUIs) for program visualization tools.

This book concentrates mainly on the second area, i.e. program visualization, with strong emphasis placed on the visualization of *parallel* programs. However, even within this simple framework it is not easy to classify some of the available visualization systems. A scientific visualization tool, for instance, may also be able to visualize data objects of a program, or a visual programming tool may permit the animation of the resulting program during execution. Nevertheless, *program* visualization will be the main topic of this book. When required, scientific visualization and visual programming will be additionally covered to point out overlapping features or differences.

2.2   Sequential and Parallel Visualization

Program visualization was already a common technique before parallel and distributed machines became widely available. BALSA (Brown, Sedgewick [38]), one of the first program animation systems, was only intended by its authors to visualize *sequential* programs. Its descendants BALSA-II (Brown [34]), ZEUS (Brown [35]) and PASTIS (Müller et al. [91]) are also only applicable to sequential programs. Consequently, all these tools can only be used to watch single processes or sequential subtasks of a parallel program. They do not have abilities to report anything about interprocess communication, load balancing, and other types of information relevant for parallel systems. New tools had to be developed, which were able to visualize the special kinds of information needed in the context of parallel computing.

The focus of this book is the visualization of *parallel* programs; however, some techniques from sequential program visualization turn out to be useful in a parallel context.

2.3 Source Code Modification

In *imperative visualization systems*, like BALSA (Brown, Sedgewick [38]), visualization is, in essence, treated as a side-effect of program execution: specific events modify a visible image in particular ways. The inherent drawback of this approach is that the user must modify the program (the source code) to be investigated.

In BALSA the user has little influence on either (i) the kind of information, or (ii) the way in which this information is displayed. The programmer has to decide a priori what is to be visualized and how. Although other program visualization systems allow the user to select from several given levels and kinds of abstractions, they usually do not permit the specification of entirely new types of abstraction. Generally, this would require identifying new events and marking them in the code. The overhead of reentering the edit–compile–execute cycle may be annoying or even prohibitive, especially when visualization is used for debugging purposes.

Possibly in response to these difficulties, a recent trend is to use *declarative* methods of algorithm visualization; thus the need for any source code modification is avoided. In yet another approach introduced by the ALADDIN system (Helttula et al. [60]), the algorithm animator defines a number of objects – usually considered to be icons – with parameters. These objects can be changed by program operation. However, as in imperative systems, the animator must still modify the program code to modify objects.

Other systems remove the need for a modification of the program code by binding object parameters to program variables. Changes of these variables are transmitted to the visualization system, which changes the items and updates the display. For instance, the PROVIDE debugging system (Moher [87]) considers all potentially interesting aspects of algorithm behavior. Procedure calls automatically inserted by the compiler inform the visualization system of all state changes.

Another method is used in BELVEDERE (Hough, Cuny [61]): modified system routines are provided which record event information in addition to performing their normal system functions. It is therefore necessary only to link the program with these special monitoring versions.

In distributed message-passing systems the declarative approach is easier to realize; a hardware or software monitor can listen to and/or record all messages during execution and pass this information on to the visualization environment.

In a shared-memory system each reference to non-local memory is a possible candidate for an event, which enlarges both the number of events and the difficulties in picking out those events which are to be animated. This may be one of the reasons why there are more tools available for message-passing computers than for shared-memory machines.

2.4   Purpose of Visualization

Several publications distinguish between the *purpose*, the *time* and the *kind of visualization* (Blaschek et al. [30], Myers [92]).

The *purpose of visualization* may be one or more of the following fields: algorithm animation, visualization of program behavior (performance), program creation, and/or program debugging.

2.4.1   Algorithm Animation

The process of *algorithm animation* is comprised of the abstraction of a program's data, operations, and semantics, and the creation of dynamic graphical views of those abstractions (Stasko [115]). Algorithm animation includes the exposition of program properties by displaying multiple dynamic views of the program and associated data structures. It also encompasses program animation and data-structure rendering, which typically involve one-to-one mappings between program data and animation images. However, algorithm animation is broader than these two areas as it involves program views that go beyond simple data-structure presentation. Frequently, the animation images have no direct correspondence to the program's data or execution units, but instead represent abstractions designed to elucidate the program semantics.

Algorithm animation systems should provide a pictorial representation of data structures used by the algorithm at a proper level of detail and abstraction. From the representation the computer user understands what the algorithms do, how they work (in a correct or incorrect way), and why they work. Multiple graphic views of an algorithm can be animated by highlighting an operation or a process with blinking, changing colors, sound, and moving objects on the screen.

The primary areas of application of algorithm animation techniques are computer science education, research in the design, development and analysis of algorithms, advanced debugging and system programming, monitoring of performance aspects, as well as documentation and description of algorithms and programs.

Systems belonging to this group are BALSA (Brown, Sedgewick [38]) and all its descendants for animating sequential algorithms and VOYEUR (Socha et al. [114]) for parallel programs.

2.4.2   Visualization of Program Behavior

Program visualization techniques give users visual feedback on the execution of a program and its processes. Graphical displays of an executed program expose some properties that might otherwise remain unnoticed. MAP

(Brewer et al. [32]) and SHMAP (Dongarra et al. [44]) animate memory references; PARAGRAPH (Heath, Etheridge [57]) provides detailed, dynamic, graphical animations of the behavior of parallel programs, and IVE (LaPolla [70, 71]) presents logical and structural software relationships, such as calling diagrams and dependency graphs. SMILI (Khanna, McMillin [66]) uses Chernoff Faces (Chernoff [40]) for visualization of program behavior. VIPS (Isoda et al. [65], Shimomura, Isoda [112]) was designed to visualize the behavior of Ada programs.

2.4.3 Visualization of Performance

Analysis of performance aspects can proceed in three directions:

Measurement: Performance measurement requires trace data from which particular performance data can be obtained. TRACEVIEW (Malony et al. [80]) displays performance aspects of parallel programs with Gantt Charts and rate displays. PARAGRAPH (Heath, Etheridge [57]) depicts processor utilization and interprocess communication. SPD (Po et al. [99]) displays utilization of each processor, memory, bus and disc. COMET (Kumar [8]) measures the inherent parallelism of large Fortran applications. To accomplish this, it neglects communication/synchronization delays, processors number limitations, etc. Although these assumptions are unrealistic, such measurements can be helpful in the design of various architectures and compilers. DPM (Miller [84]) measures communication statistics, provides causality analyses[2], and measures the parallel speed-up. MATRIX (Paul, Poplawski [97]) can be used for performance evaluation of parallel algorithms for dense matrix operations. Interactive performance analysis similar to the use of interactive debuggers is accomplished with PPUTTS (Fowler et al. [49]). With PATOP, part of the TOPSYS project (Baker et al. [27]), system resources can be visualized in relation to some programming units and isolated from them.

Modelling: With PAW (Melamed, Morris [83]) and its successor Q+ (Funka-Lea et al. [52]), it is possible to create a queuing network for simulation studies of system performance. The goal of the PARET project (Nichols, Edmark [95]) is the design and study of multicomputers as systems, rather than as isolated components. Multicomputer subsystems are represented as directed flow graphs.

---

[2]The basic strategy for a *causality analysis* is to identify each request to a given processor and to follow the sequence of interactions within the processor caused by that request.

Tuning: Performance tuning tools help the analyst in detecting performance bottlenecks. TANGO (Stasko [115], Davis et al. [42]) determines
and shows resource requirements of (possibly unavailable) hardware resources. PIE (Lehr et al. [75]) provides ways to observe how computations are executed. TASKGRAPHER (El-Rewini, Lewis [47]) is a tool for
scheduling parallel programs with different scheduling algorithms. IPS-2
(Miller et al. [85]) provides performance analysis techniques that automatically guide the programmer to the location of program bottlenecks.

2.4.4 Program Creation

Program Creation is essentially what is called *visual programming* in Section 2.1. This topic is beyond the scope of this book. Related articles can
be found in Brown et al. [124] and Yau, Jia [125].

Systems that also do some kind of program creation aside from program
visualization are FAUST (Guarna et al. [54]), POKER (Snyder [113]), PARET
(Nichols, Edmark [95]), Q+ (Funka-Lea et al. [52]), PAWS (Pease et al. [98]),
PECAN (Reiss [103]), PEGASYS (Moriconi, Hare [90]) and TOPSYS (Baker et al.
[27]).

2.4.5 Debugging

An increasing number of systems and tools use visualization for debugging purposes. VIPS (Isoda et al. [65], Shimomura, Isoda [112]) is a debugging tool for
Ada programs. It provides graphical views of data values, data-flows, and the
calling sequence of subprograms. PROVIDE (Moher [87]) offers a continuous
display of user-defined process state information and random access to all process states during program execution; VISTA (Tuchman et al. [118]) shows all
messages sent between processes, thus helping in debugging. BUGNET (Wittie
[122]), a debugger for distributed C programs running on UNIX machines,
gives information about interprocess communication, I/O events, and execution traces for each process. The unique feature of BELVEDERE (Hough, Cuny
[61]) is its ability to view the system behavior from user defined *perspectives*
during debugging, which is an attempt to compensate for the asynchronous
behavior of the parallel program.[3] The POKER environment (Snyder [113]) allows the modelling of communication graphs; processor variables are displayed
in boxes connected by communication links.

---

[3]Events belonging to the same logical system-wide event, such as when all processes send
a value in a common direction, are shown together even if they do not occur at exactly the
same time.

2.5 Time of Visualization

Visualization takes places either (i) during the execution of the program (*real-time visualization*) or (ii) after its execution (*post-mortem visualization*).

2.5.1 Real-Time Visualization

Real-time visualization permits very accurate simulations[4] of the interaction of processes. However, it may be impossible to recreate some situations of interest. One approach to achieving real-time visualization is to send information to an animation tool at breakpoints, as is done in PASTIS (Müller et al. [91]). Another approach, used in ZEUS (Brown [35]), automatically displays views based on events generated by the algorithm.

2.5.2 Post-mortem Visualization

Post-mortem visualization requires some sort of *trace file* which is recorded during the program execution. Difficulties arise in assuring accuracy. Because of the nondeterminism of parallel programs, the execution path, i.e. the relative order of the recorded events, is essential. Some timing values (*time stamps*) for each operation must be generated by the trace file generator in order to guarantee this order. Hough, Cuny [63] present a technique for "reordering" events without violating dependencies and thereby producing equivalent visualizations.

2.6 Kind of Visualization

Any visualization illustrates one (or more) of the following aspects: algorithm, interconnection topology, mapping, data structures, program code, program behavior, and/or performance.

2.6.1 Visualization of Algorithms

Algorithm visualization consists of animated pictures based on program events generated by the algorithm that is visualized. In particular the earliest visualization systems, like BALSA (Brown, Sedgewick [38]) and ZEUS (Brown [35]) were intended for algorithm visualization.

---

[4]The term "simulation" is used here because the execution of a program may be disturbed by monitoring or visualization statements (due to the *probe effect*).

2.6.2   Interconnection Topology

In some tools, such as TASKGRAPHER (El-Rewini, Lewis [47]), a graph editor
is used for designing *task graphs* which are descriptions of the hardware topology. The task graph is then used as a basis for performance analysis of that
particular interconnection topology, or can even be used to predict the performance of any conceptual topology. However, the task graph is not used for
visualizing the topology itself. A real animation of the interconnection topology
in various layouts is provided in PARAGRAPH (Heath, Etheridge [57]), where
pictorial representations of nodes (processors) change their color according to
their current state.[5] POKER (Snyder [113]) lets the user define the communication structure of an algorithm graphically by embedding it in a two-dimensional
lattice. Sequential code, programmed in a conventional programming language
(Pascal, C, etc.), may be assigned to individual processors to watch execution
behavior.

2.6.3   Mapping

Mapping refers to the distribution of pieces of work associated with a parallel
program among the available processors. In IPS-2 (Miller et al. [85]) a machine-
graph containing numbers and types of processors and interconnections can be
drawn. This graph can then be used to assign each procedure to any one of the
processors (procedures are displays hierarchically) and to visualize different
program aspects, like execution time, communication time or load balance.
PAWS (Pease et al. [98]) permits the evaluation of conceptual machines before
the building of any specific hardware by translating applications written in a
high-level source language into a machine-independent data-dependence graph
which is heuristically mapped onto different target architectures.

2.6.4   Data Structures

There is a great deal of scientific visualization software available which supports
the visualization of large amounts of data irrespective of their origin. CLAM and
CLAMSHELL (Foulser, Gropp [48]), for instance, support visualization of large
amounts of (statistical) data (statically or animated 2D and 3D plots, surface
and contour plots). IVT and DT (Gupta et al. [20]) help in understanding
neural networks by visualizing the properties of each neuron (distribution of
weights, activation values, etc.) in different levels of detail.

There are also tools, however, which present program variables and structures in a graphical way, mainly for debugging purposes. VISTA (Tuchman et al.

---

[5]The current version of PARAGRAPH supports layouts for hypercubes, rings, meshes,
trees, butterflies, crossbars, and several other interconnection topologies. Even user-defined
layouts are possible.

[118]) constantly shows program variables during execution. VIPS (Isoda et al. [65], Shimomura, Isoda [112]) and Pv (Brown et al. [124]) display dynamic data structures (trees, queues, etc.) as graphical objects (nodes) that are connected with lines or arrows (pointers). The user can choose the level of detail. Value updates, creation, rearrangement and deletion of nodes are visualized. MATRIX (Paul, Poplawski [97]) shows which part of a two-dimensional matrix is being operated on by which processors at what time. PASTIS (Müller et al. [91]) visualizes program structures which are retrieved from gdb, the GNU source level debugger. POKER (Snyder [113]) offers the visualization of program variables; furthermore, the values of these variables can be changed during the execution, i.e. the new values are written back into the processor memories.

2.6.5 Program Code

Code visualization shows the individual statements of a program. In BALSA (Brown, Sedgewick [38]) the current execution point is shown by highlighting of the corresponding line in the source code and the procedure hierarchy is displayed in overlapping windows. The FAUST environment (Guarna et al. [54]) depicts the entry and exit points of each routine with distinct symbols and shows those symbols on a time-line diagram. PEGASYS (Moriconi, Hare [90]) displays pictures composed of icons and their properties (size, location, etc.). Icons in the picture correspond to a computational concept expressed by the picture (for instance, process, communication line or data package).

2.6.6 Program Behavior

Tools of this category attempt to give detailed insights into the internal course of events taking place in parallel programs. PIE (Lehr et al. [75], Rudolph, Segall [111]) and MKM (Lehr et al. [73, 74]) show the state of processes and processors in a time diagram. BELVEDERE (Hough, Cuny [61, 62]) allows the description, manipulation and animation of logically structured patterns of process interactions. BUGNET (Wittie [122]) displays information about the message traffic between processes. PARAGRAPH (Heath, Etheridge [57]) displays numerous aspects of program behavior: processor status, load balance, message queues, and current tasks to mention only a few. The main goal of PPUTTS (Fowler et al. [49]) is to understand the behavior of incorrect or inefficient parallel programs on large-scale, shared-memory multiprocessors. VOYEUR (Socha et al. [114]) offers facilities to construct application specific views for parallel programs, ranging from simple textual representations to complex graphical maps. These views help to simplify the exploration of program behavior and detection of bugs. Q+ (Funka-Lea et al. [52]) describes a model by drawing a picture of it. Thus system behavior can be observed via the animated movement of traffic entities within the model and the gradual evolution of statistics.

2.6.7   Performance

PREFACE (Bernstein et al. [29]) allows the visualization of four different categories (user work, system work, idle time, and different types of overhead). SPD (Po et al. [99]) uses bar charts to display processor, memory, bus, and disk utilization. The general purpose trace visualization system TRACEVIEW (Malony et al. [80]) provides Gantt Charts and rate displays for measuring I/O and CPU memory references. COMET (Kumar [8]) measures the total possible parallelism of Fortran programs. IPS (Miller, Yang [86]) provides critical path and phase behavior analysis[6]. SIGMA, part of the FAUST environment (Guarna et al. [54]), helps in optimizing the application code by answering questions such as where a variable was initialized, which routines modify or use a variable, what possible side-effects of a routine are, whether a particular loop can be parallelized or vectorized, and – if not – which variables prohibit concurrency. PARAGRAPH (Heath, Etheridge [57]) offers utilization Gantt Charts, concurrency profiles and critical path analysis. SMILI (Khanna, McMillin [66]) uses Chernoff Faces for performance evaluation . TASKGRAPHER (El-Rewini, Lewis [47]) offers speed-up line graphs, a Gantt Chart scheduler, processor utilization and efficiency charts, and critical path displays for performance tuning . PATOP, part of the TOPSYS project (Baker et al. [27]), uses histograms and bargraphs for performance monitoring.

---

[6]The goal of the *phase behavior analysis* is to identify phases in a program's execution history. Each phase represents a simpler subproblem with specific execution characteristics.

Chapter 3

Design Goals and Techniques

3.1 Design Goals

The principle goals in the design of performance visualization software are *ease of understanding*, *ease of use*, and *portability* (Heath, Etheridge [57]).

3.1.1 Ease of Understanding

Since the key point of visualization is to facilitate human understanding, the visual display should be as intuitively meaningful as possible. Graphical elements symbolizing data attributes should be effective and expressive. Display techniques should take advantage of well known and common metaphors from daily life. They should use existing conventions to support the human visual system.

3.1.2 Ease of Use

Dynamic high-level graphical interfaces should significantly increase the ease of use of visualization software. Nowadays interactive, mouse and menu-oriented user-interfaces allow the user to interact with dynamically changing graphical representations of data and programs. Additionally, the operating system should support multiple windowing and multitasking techniques. Another important factor is that the program under study need not be modified extensively to obtain visualization data. If performance analysis tools are difficult to use, software developers will reject them in favor of simpler, inferior techniques. Also, if the performance instrumentation environment does not permit various approaches to performance data reduction and analysis, its functional lifetime will be limited.

Ease of use should also cover the aspect of extendibility. Users should be able to construct application specific views of parallel programs and should also have possibilities to add new displays of their own design.

3.1.3   Portability

A visualization system should not depend on particular properties of the target hardware or system software, i.e. it should be portable to the greatest possible extent. The visualization system should be capable of displaying the performance data of different parallel architectures and parallel programming paradigms. Existing multiprocessor machines range from coarse grain parallelism, where a few processors execute different instructions on different data (MIMD machines, like the Cray Y/MP), to fine grain parallelism, where a great number of processors execute the same instructions on different data (SIMD machines, like the Connection Machine with up to 65536 processors). Parallel hardware may be tightly coupled (via a shared memory) or loosely coupled (with message passing).

3.2   Demands and Suggestions

Creating and using an algorithm animation system to get efficient dynamic visualizations of computer programs appears difficult; however, in practice, successful algorithm animations make use of only a small "bag of tricks" as discussed in the following section (Brown, Hershberger [36], Brown, Sedgewick [39]).

3.2.1   Multiple Views

A fundamental thesis of any algorithm animation system is that a single view of an algorithm or data structure does not tell its complete story. It is generally more effective to illustrate an algorithm with several different views. Even for showing simple algorithms, such as Quicksort, in multiple aspects, a single view must encode so much information that it becomes difficult for the user to pick out the details of interest from the wealth of information on the screen. The approach of using multiple views is advantageous because each individual view displays a small, comprehensible aspect of the algorithm. The composition of several views is more informative than the sum of their individual contributions.

3.2.2   State Cues

Changes in the state of an algorithm's data structures are reflected on the screen by changes in their graphical representations. For example, a node of a tree is highlighted while it is being processed. State cues can further be used to link different views together (a single object is represented the same way in every view where it appears) and to reflect the dynamic behavior of an algorithm.

3.2.3   Static History

Static histories enable the user of an animation system to become familiar with the dynamic behavior of an algorithm at his own speed. He can focus on crucial events where significant changes happen, without paying too much attention to repetitive events.

3.2.4   Amount of Input Data

It is important to start the animation with a small problem instance, preferably with textual annotations, to relate the visual display to the user's previous understanding. Once the user has established the connection between the textual and the graphical representation of the algorithm, he can omit the labels and proceed to larger, more interesting data sets.

3.2.5   Choosing Proper Input Data

The choice of input data strongly influences the message that an animation conveys. It is often instructive to use *pathological data* to push the algorithm to an extreme behavior. Another example of choosing data for pedagogical purposes is to use some kind of data filtering. If, for instance, a hashing algorithm is so effective that rehashing almost never occurs in practice, the user can filter out some of the randomness in the input data to make the animation more interesting.

3.2.6   Color Techniques

Color may be used to achieve one (or more) of the following goals:
- reveal an algorithm's state (by mapping states to colors);
- unite multiple views (objects are displayed uniformly in each view);
- highlight areas of interest;
- emphasize algorithmic patterns; and
- make the course of computation visible (for instance, by mapping time intervals to colors).

3.2.7   Audio Techniques

Auralization, also called sonification, might enhance visualization in the following areas:
- auralization reinforces visual views (each action produces a certain tone);
- audible information conveys patterns ("hearing" an algorithm can reveal distinctive signatures);
- audio cues signal exceptional conditions (like warnings, etc.); and

- auralization may replace simple visual views (to save screen space for other types of visual views).

3.2.8   Scripts

Every convenient environment should include a facility for saving and invoking sequences of operations, called *scripts*. In BALSA (Brown, Sedgewick [38]) scripting facilities are quite rudimentary: everything the user does is saved in some file to be played back later. Some features exist that allow different things to happen during the playback: pausing the script, i.e. waiting for the user to press a key, or saving complex window configurations. It is difficult, though possible, to edit BALSA scripts.

More sophisticated scripting facilities are implemented, for instance, in BALSA-II (Brown [34]). Rather than recording keystrokes and mouse movements a BALSA-II script stores higher level information. Thus such a script can be more complex, but can still easily be edited after creation.

3.3   Role of the Environment

A software development environment for performance tuning or debugging is not effective if it demands an inordinate amount of effort from its users to search for performance degradation problems, especially for problems that could easily be revealed by automatic techniques (Francioni et al. [50]). On the other hand, an environment that makes qualitative judgements about a computation's performance without the user's permission may annoy and mislead the user. An environment should present the information it retrieves about computations in forms that assist the user in making his own qualitative judgements about how his computations behave.

3.4   Handling Information

Parallel program analysis is essentially a problem of information management (Lehr et al. [75]). The amount of execution states that must be examined is overwhelming; therefore, flexible techniques for selecting and presenting information are crucial. The user of a tool must be supplied with a complete picture of a program's execution. Information must be available about all aspects of the program's behavior, and the user should be able to view this information in different levels of abstraction. The system must supply this large body of performance information in such a way that the user is not overwhelmed, but can easily and intuitively access the information. In addition, the user needs more than a tool that provides extensive lists of performance metrics; he needs one that will direct him to the location of performance problems. One way to accomplish this is *critical path analysis*, which is used, for instance, in IPS (Miller, Yang [86]), PARAGRAPH (Heath, Etheridge [57]) and TASKGRAPHER (El-Rewini, Lewis [47]). Critical path analysis displays the longest serial thread in the parallel computation. This is the primary place to look for potential algorithm improvements. Another approach is realized in the FAUST environment (Guarna et al. [54]), in which the user can ask the system where a variable was initialized, updated or used.

3.5 Levels of Analysis

Some authors suggest that an environment should support multiple analysis levels (Rover, Wright [109]), including at least the following perspectives:
- the program, or application;
- the architecture, or logical network; and
- the machine, or physical network perspective.

It should be possible to dynamically configure data analysis and presentation components, and thus allow the performance analyst to change perspectives during execution. This would be accomplished by offering multiple views and enabling the user to choose among these different views at any time during the program execution.

3.6 Continuous Display

Using a traditional debugger the user is compelled to ask the system repeatedly for the value of a particular program object. Moher [87] suggests the use of *continuous displays* with which the user needs to specify the object of interest only once. The displays then allocate a permanent screen area and are automatically updated to reflect the ongoing course of computation.

3.7 Measurable Parameters

A variety of parameters and summary statistics can be collected and then supplied to the user via animations (Rover [106]):
- operation count (total number of operations so far);
- current work (number of processes/processors currently running);
- computation time;
- execution rate (operation count per time interval);
- message traffic (number of messages or number of bytes sent);
- communication time (absolute, or as a fraction of the computation time);
- wait time (idle time, when a process waits for a message);
- communication overhead (data transfer between processors);

- granularity factor (many tiny parallel program chunks or a few large);
- activity distribution (work of each processor over time and space);
- processor concurrency and utilization (for a subset or all processors);
- computational balance (ideally, equal work for each processor); and
- communication balance (among particular processors or over time).

3.8   Visualization Primitives

Mapping numbers to symbols to enhance the ability of human interpretation is well known as a useful technique. Graphical elements are the building blocks for complex displays (Domik [18]). These elements are referred to as *visualization primitives* and include the following: length or size, position (usually 3 dimensions), angle, slope, area, density, color saturation and color hue, texture, and shape.

3.9   Display Types

There are a great deal of display types that have been used in performance visualization:

Meters represent one-dimensional information, usually some kind of state information. Meters can be distinguished in the following way: *analog meters*, like a conventional clock; *digital meters*, used to supply accurate values; *move meters* (Figure 5.4) and *zoom meters*, both suitable for quick interpretation of the represented data; and *symbol meters* displaying threshold values.

2D Plots, also called *X-Y* plots, found acceptance long before the term visualization became popular. *Line plots* (Figure 5.9) are used for activity profiles, execution speed-up and memory references, while *scatter plots* fit best for displaying density distributions. *Surface* and *contour plots* highlight activities within a given data set. *Gantt Charts* (Figure 5.2, 5.14), plots with time on either axis, show performance changes over time. *Histograms* (Figure 5.3, 5.5) and *Leds* (discrete histograms, Figure 5.10, 5.13) are capable of displaying single values for a number of instances.

3D Plots are used to emphasize dependencies among performance values.

Percentage Profiles display one-dimensional information either in a linear fashion, called a *bar diagram* (Figure 5.16), or within a circle, called a *pie chart*.

Chernoff Faces (Chernoff [40]) map multivariant data to cartoon-like visualizations.

Kiviat Diagrams (Kolence, Kiviat [68]; Figure 5.6) are used to reveal certain kinds of data distribution. In this type of diagram there is an axis for each data dimension, starting from a center of a circle to its border. Points near the border represent maximum value, near the center minimum value.

Graphs (Figure 5.12) contain nodes, connections between the nodes and labels. They can visualize certain dependencies of program data, code, communication or events.

Matrix Displays (Figure 5.11) usually display a two-dimensional array of data. Each cell of the matrix is filled with a particular color or texture according to the current value.

Chapter 4

Available Software

This chapter gives a more detailed description of some selected pieces of available software. Rather than listing all systems that might be similar in many aspects, this chapter provides information about some different directions that tool developers have taken.

4.1 BALSA

The common approach today for animating algorithms specified in high-level procedural languages was pioneered in BALSA — Brown Algorithm Simulator and Animator (Brown, Sedgewick [38]). BALSA is a special-purpose system tailored for performance investigations (in particular for probing the behavior of programs in action).

Programs are augmented with user-defined *interesting events*, which are supposed to cause changes of the displayed image, called *view*. Events have optional parameters, which typically specify the event more precisely.

Each view controls a window on the screen which is notified when one of the specified events happen in the algorithm. A view is responsible for updating its graphical display appropriately based on the incoming event information. Views can also return information from the user to the algorithm.

The original version of BALSA was implemented on Apollo workstations. It was designed for the visualization of *sequential* algorithms, and is accordingly only second choice for visualizing of parallel algorithms (Section 2.2). Today BALSA is available on a large variety of hardware platforms, including PCs and Macintoshes.

4.1.1 Animation of an Algorithm

There are several types of activities associated with BALSA:

1. The BALSA facilities may be used to get a dynamic, graphical display of the program in execution. Decisions have to be made what kind of information should be visualized, and *interesting event* signals have to be included in the algorithm.

2. BALSA's low-level facilities may be used to display programs in execution. Software that creates and animates graphical images has to be written based on the interesting events chosen. This results in a set of algorithms and views which are accessible to the user.

3. BALSA's high-level facilities may be used to assemble algorithms and views into coherent entities and to present existing views to the user. To accomplish this, *scripts* (lists of recorded actions) must be written.

4.1.2   User Facilities

The user of BALSA has three different types of capabilities:

1. *Display primitives* manipulate the presentation of views. The user can create, size and position any "view window" by using a mouse on pop-up menus and other standard window operations.

2. *Interpretative primitives* control the execution of the program: starting, stopping, slowing down, and running the algorithm backwards. Other supported features are breakpoints and stepping through the program in units meaningful to the algorithm.

3. *Shell primitives* allow the user to save and restore configurations, and to save and invoke *scripts* consisting of sequences of BALSA primitive operations.

4.2   Descendants of BALSA

Tools directly influenced by BALSA are BALSA-II, ZEUS and PASTIS.

4.2.1   BALSA-II

BALSA-II (Brown [34]) deals with a program in three parts:

1. the *algorithm* itself, which, again, must be augmented with interesting events;

2. various *input generators*, that provide the necessary input data for the algorithm to be visualized; and

3. a set of *views* that present animated pictures of the algorithm in action.

In practice, neither input generators nor views need to be implemented for each algorithm because existing components from a standard library can be used.

BALSA-II provides scripting facilities (Section 3.2.8): instead of recording every keystroke or mouse activity, higher level information is stored, enabling the user to switch between passive viewing and active exploration. After interrupting the running script, the user may invoke an arbitrary number of actions, and then resume the script at the point it was interrupted.

4.2.2 ZEUS

A novel feature of ZEUS (Brown [35]) is its automatic generation of utility views based on the set of interesting events of the investigated algorithm. This is accomplished with ZEUS' preprocessor ZUME, which is also responsible for dispatching events between algorithm and views. Two of ZEUS' utilities are particularly noteworthy:

1. The *transcript view* contains a typescript that displays each generated event as a symbolic expression.

2. The *control panel* contains a button for each event, with appropriate widgets for specifying additional parameters. Clicking on any one of these button causes the respective event to be generated.

These utilities are extremely valuable for debugging algorithms as well as new views. With these two views it is possible to develop and debug a view before a specific algorithm is investigated.

4.2.3 PASTIS

PASTIS (Müller et al. [91]) does not require any change of the source code. This is made possible by the use of an existing source code debugger, such as the GNU source level debugger **gdb**, which receives program data and sends it to the animation module of PASTIS. With the possibility to set breakpoints in different ways and query the value of variables and expressions at these breakpoints, there exist numerous possibilities for algorithm animation.

The visualization model used by PASTIS is an object oriented extension of the following relational data model: *visualization scripts* have parameters which are bound to data values extracted from the program. The class of an animation script is specified by its parameters, whose format is described in terms of relations.

Visualization scripts describe how source data is to be mapped into animations. The first part of the visualization script declares the interface of the *animation scripts* used in the visualization. In the second part of the visualization script the animation scripts are defined and the way they are supplied with data is declared. The visualization script is read by the process executing the given program under the control of the debugger, which sends program data and additional control statements to the animation server. The animation server, running as a separate process, distributes the incoming data to the individual animations[1]. The user has the choice of taking an animation script from a library, or he can write a new animation script with the support of the animation editor.

PASTIS offers three classes of animation scripts:

1. A *tuple* is composed of values of an *elementary data type.*[2]

2. A *relation* is a set of tuples of the same format. To identify individual elements of a relation, *keys* are introduced. Any key can be defined by one or more entries of the relation, where each entry must uniquely identify the elements of the relation.

3. A *network* defines relationships between the elements of one or different relations. It depicts the relationship between individual elements, like the display of a linked list.

The graphical representation of PASTIS is built on the X window system, where each animation is an X-client. The user can influence the form of the presentation with three types of commands:

1. *Viewing commands* permit zooming and scrolling.

2. *Information commands* allow the user to obtain textual information on data hidden behind an item.

3. *State control commands* include iconification, the stopping or killing of animations and the initiation of sub-animations.

4.3   PARAGRAPH

PARAGRAPH (Heath, Etheridge [57]) is a graphical display system for visualizing the behavior and performance of parallel programs running on message-passing multiprocessor machines. It takes as input execution trace data provided by PICL (Portable Instrumented Communication Library, Section 9.1), developed at Oak Ridge National Laboratory. Besides providing communication functionality PICL optionally produces an execution trace of a parallel

---

[1]Animations are animation scripts in execution.

[2]PASTIS' elementary data types are: `integer`, `real`, `boolean`, `string` and `animation`.

program running on a message-passing machine. The resulting trace data can then be replayed pictorially with PARAGRAPH to obtain a dynamic, graphical depiction of the behavior of the parallel program. PARAGRAPH provides several distinct ways to view processor utilization, communication traffic, and other performance data to enable insights that might be missed otherwise by examining just a single view.

PARAGRAPH is based on the X Window System and runs on a wide variety of workstations. Although PARAGRAPH is most effective in color, it also works on monochrome and grayscale monitors. It has a graphical, menu-oriented user interface that accepts user input via mouse clicks and keystrokes. The execution of PARAGRAPH is event driven, including both user-generated X Window events and trace events in the input data file provided by PICL. Thus, PARAGRAPH displays a dynamic depiction of the parallel program while also providing responsive interaction with the user. Menu selections determine the execution behavior of PARAGRAPH both statically (i.e. initial selection of parameter values) and dynamically (i.e. pause/resume, single-step mode). PARAGRAPH preprocesses the input trace file to determine relevant parameters (i.e. time scale, number of processors) automatically before the graphical simulation begins, but these values can be overridden by the user, if desired.

PARAGRAPH currently provides more than 25 different displays or views, all based on the same underlying trace data, but each giving a distinct perspective. Some of these displays change dynamically in place, with execution time in the original run represented by simulation time in the replay. Other displays represent execution time in the original run by one space dimension on the screen. The latter displays scroll as necessary (by a user-controllable amount) as visual simulation time progresses. The user can view as many of the displays simultaneously as will fit on the screen, and all visible windows are updated appropriately as the trace file is read. The displays can be resized within reasonable bounds. Most of the displays depict up to 512 processors in the current implementation, although a few are limited to 128 processors.

PARAGRAPH is extendible so that users can add new displays of their own design that can be viewed along with those views already provided. This capability is intended primarily to support application-specific displays that augment the insight gained from the generic views provided by PARAGRAPH. Sample application-specific displays are supplied with the source code. If no user-supplied display is desired, then dummy "stub" routines are linked with PARAGRAPH instead.

The PARAGRAPH source code comes with several sample trace files for use in demonstrating the package and verifying its correct installation. To create individual trace files for viewing with PARAGRAPH, PICL is also required. The tracing option of PICL produces a trace file with records in node order. For graphical animation with PARAGRAPH, the trace file needs to be sorted into time order, which can be accomplished with the UNIX sort command.

By default, PARAGRAPH automatically detects the appropriate display mode (color, grayscale, or monochrome), but a particular display mode can be forced, if desired, by the corresponding command-line option. This facility is useful, for example, in making black-and-white hardcopies from a color monitor. PARAGRAPH optionally uses several environment files to customize its appearance and behavior. In this way (i) the initial state of PARAGRAPH upon invocation (i.e. open windows/displays and various options), (ii) the order and alternative names for the processors, and (iii) the color scheme for identifying tasks may be specified.

Chapter 5 contains a detailed description of PARAGRAPH including illustrating figures for most of the available displays.

PARAGRAPH is a public domain software product. It can be downloaded from netlib (netlib@ornl.gov) or via anonymous ftp (ftp.ncsa.uiuc.edu).

4.4 TRACEVIEW

TRACEVIEW (Malony et al. [80]) is designed as a general-purpose trace-visualization system. On the one hand it is flexible enough to let the user select analysis and display alternatives; on the other hand, it provides a structure rigid enough to build on the resources of the tool user-defined facilities and extend the basic analysis and display methods. This is accomplished by the use of a simple interpretation model. Of course, TRACEVIEW cannot support all trace-visualization models, but it does support frequently occurring, simple visualization problems, and provides extension mechanisms for customizing more complex problems. Its graphical interface is based on the Motif widget set.

The concept of TRACEVIEW is based on a trace-visualization session, which is the root of a hierarchical tree structure dealing with trace files, views and displays. A session is related with one or more trace files. For each trace file the user can define a set of views. A view defines a trace subregion by specifying a particular event filtering and a time interval. For each view the user can specify a set of displays that combine data from multiple traces.

4.4.1 Session Management

The session manager coordinates the trace, view and display managers and is able to save and restore the current session, which is defined as the set of open trace files, the set of opened views for each trace, and the set of displays for each view. The current session can also be extended by merging it with other, previously defined sessions (including trace files, views and displays); existing conflicts in the trace information are automatically detected.

To the user, the session manager appears as a large window, containing three subwindows (files, views, and displays window).

4.4.2 Trace Management

The trace file is divided into two parts: an ASCII header and binary trace data. The header specifies the number of data fields, how each data field is labelled when presented to the user, the total number of events and the total time represented by the trace data.

The trace data consists of a time-sequenced list of trace events. Each event reflects the instance of some action taking place during the investigated computation. TRACEVIEW makes no semantic interpretation of the actions that the event represents; it only interprets them as state transitions. Each event recorded in the trace file includes an event type, a time stamp and information about the state being exited and the state being entered. Additional data fields may be included when required. Events within a trace file are in a homogeneous format; this means that each event has a fixed number of data fields associated with it.

The trace manager is responsible for opening and closing trace files, interpreting the trace file header, calculating global trace statistics and reading events from open trace files.

4.4.3 View Management

A view definition specifies a starting time and a final time in the trace, and a list of names of events to be filtered from the trace. The user defines every view in a separate window. This window contains the name of the view, the name of the trace file concerned, a list of all defined events (where each event can be selected/unselected either individually or by using string pattern-matching capabilities), and the range selection for both the lower and the upper bound (for each either the time and/or an event can be specified).

Based on the user's selections TRACEVIEW transforms raw trace information into a virtual trace.

4.4.4 Display Management

The display manager offers a list of existing displays to select for the view. The user can open, create and omit displays. The display manager does not actually display any data; rather it invokes a display module to display the virtual trace. TRACEVIEW does not dictate a specific display method: although two display types are statically linked with TRACEVIEW, the user can incorporate any display method able to interpret a TRACEVIEW virtual trace. TRACEVIEW provides two built-in display types:

Gantt Display is used for displays based on state transitions, where the x axis represents time, and the y axis varies from chart to chart. For traces that contain no data fields, the display manager shows only a single chart with states represented on the y axis. If the trace also includes data fields, a Gantt Chart for each data field is displayed.

Rates Display depicts the number of times a state is entered. Rates displays are only defined if the trace includes data fields, in which case the x axis represents ordinal instances of the states being entered and the y axis represents the individual data field values. A chart is available for each data field defined in the trace; additionally, the summation of all the metrics is provided by the display manager.

Both display types use the *Gantt Chart widget*, which was specially developed for TRACEVIEW. A key feature of this widget is its ability to display a huge amount of data. Usually the number of points to display greatly exceeds the pixel width of the Gantt Chart's x axis. This widget offers two optional solutions: density bars and average curves.

Density Bar (displayed above the data display area) is a band of colors. The user can choose a color map, which assigns integer ranges to colors. In a *value-density bar*, the color at each pixel represents the average of all the data points displayed at that pixel on the x axis. In a *point-density bar*, the color at each pixel represents the number of data points at that pixel on the x axis.

Average Curve overlays the data displays. It takes the average of a fixed interval of points as the interval slides along the x axis with the current pixel in the center of the interval.

4.5   MAP and SHMAP

The Memory Access Pattern (MAP) program (Brewer et al. [32]) and the Shared Memory Access Pattern (SHMAP) program (Dongarra et al. [44]) are tools for the software development on high-performance SIMD and MIMD machines, respectively. Both tools provide a graphical display of memory access patterns in Fortran algorithms to help detect system bottlenecks. The memory hierarchy, which can be visualized, takes the following form: main memory, cache, local memory, and vector registers.

Animating with these tools requires a preprocessor, called (Shared) Memory Access Pattern Instrumentation MAPI/SHMAPI, which analyzes an arbitrary Fortran program and generates a call to one of the (Shared) Memory Access Pattern Animation MAPA/SHMAPA routines for each reference to a matrix element. Also references to the Level 1, 2, or 3 BLAS (Lawson et al. [199],

Dongarra et al. [197, 198]) routines are translated to MAPA/SHMAPA routines. The replaced routine records the memory access to be made, as well as the number of floating point operations to be performed, and then calls the original BLAS routine. This approach reduces the size of the trace file dramatically. The output of MAPI/SHMAPI is a Fortran program, which when executed and linked with the MAPA/SHMAPA library, executes the original code and generates an ASCII trace file containing three kinds of trace events: array definition, read access, and write access. Once a trace file has been created, it may be used with MAPA/SHMAPA in analyzing a given algorithm by using several different system configurations. Thus MAP/SHMAP provides insights into the operation of the algorithm on various systems.

4.6 PIE

The PIE system (Rudolph, Segall [111]) offers a framework for the development of efficient parallel software. Many efforts are made to predict, detect, and avoid performance degradation. It consists of three components: the modular programming metalanguage, the program constructor and the implementation assistant.

4.6.1 Modular Programming Metalanguage

The goals of the modular programming (MP) metalanguage are to provide support for the efficient manipulation of parallel modules, fast parallel access to shared data constructs, and programming for observability in a language-independent fashion. The MP metalanguage allows programmers to specify most of the code in their programming language of choice. It provides a set of modular constructs that encapsulate the elements of a parallel program:

Activity is a sequence of sequential operations. An activity is considered the smallest schedulable unit of sequential work. The activity construct includes the set of local declarations, local data, the activity body written in a sequential language, *sensors* (see below), and monitoring directives associated with activity functions. Sequential languages are extended by MP metalanguage synchronization, initiation of parallel operations, and access to shared data.

Frames specify access and management of shared memory. A frame consists of declarations of shared data, a set of operations on the shared data, constraints on the execution of these operations, and monitoring directives associated with the shared data and its access. The operations of a frame are automatically unfolded by the MP preprocessor into the activity code. Coordination and synchronization, both required for correct parallel access to shared memory, are specified by constraints on the parallel execution of these operations. These constraints specify the required access sequences as well as the parallel actions that are forbidden.

Team is a set of activities to be executed in parallel, a set of frames, and control information; moreover teams specify monitoring directives for the whole module as well as resource allocation (dependent on the resources supplied by the target machine). MP metalanguage enforces scoping in a manner similar to that found in block-structured sequential programming languages.[3]

Sensors provide the basic mechanism to observe or monitor the execution of the program. Sensor implementation is automatic and may be in hardware or software.[4] Since PIE does all the bookkeeping associated with monitoring, the programmer can also see the surrounding code and specifications when viewing the result of a particular sensor. MP supports four types of sensors:

1. time consumed by a team or activity;

2. time required to execute a code block;

3. the value of a specified variable; and

4. user-defined sensors.

Sensors have the following main characteristics:

1. they are able to detect a precisely defined event of different types (see above);

2. they can cause a transfer of information from the program to an external environment in an automatic and controlled way;

3. they are transparent to the correctness and (as far as possible) to the performance of the program; and

4. they can be enabled and disabled under the control of an external mechanism allowing customization of monitoring requirements at run time.

4.6.2   The Program Constructor

The program constructor provides high level support for the generation of efficient parallel programs. It builds on MP's concept and frees the programmer

---

[3]Each logical entity can reference other entities either in the same team or in an outer team.

[4]The current version of PIE supports only software sensors.

from much of the specification. This is realized by providing mechanisms and policies for combining and transforming both static and dynamic information into meaningful program representations based on a relational model approach. The PIE program constructor consists of three main elements:

MP Oriented Editor presents a partially constructed parse tree, which consists of nonterminal and terminal nodes, to the programmer. When program information is entered nonterminal nodes are expanded, the parse tree is refined, and some semantic action affecting the contents of the relational model may occur.

Status and Performance Monitor provides significant information to programmers on the late phase of program development. The monitoring environment is composed of sensors, monitoring directives (specified throughout MP specifications) and presentations of the monitored data (views can be displayed after the parallel program is executed).

Relational Representation System makes multiple representations of the syntactic and semantic information available both at development time and run time. Modules defined in MP – activity, frame and team – are directly visible and can be manipulated through the MP oriented editor. Specific views are obtained by applying relational operations to PIE's relational model. Examples of representation are communication graphs/trees, activity graphs/trees, activity-resource graphs/trees, and dependency graphs/trees. The graph representations are for the development phase and the trees refer to run time representations.

4.6.3 The Implementation Assistant

The goals of the implementation assistant are to provide semantic support for the parallel program development cycle in the following ways:

1. predict parallel program performance before extensive implementation investments;

2. assist the user in choosing implementations;

3. generate parallel programs semiautomatically by supporting a set of well-defined implementations;

4. support semiautomatic semantic instrumentation; and

5. support a set of predefined representations related to the provided implementations.

An implementation is a specific way to decompose the parallel computation activities and data, as well as the way they are controlled. Available possibilities are *Master-Slave* (one master and a set of slaves), *Recursive Master-Slave* (same as Master-Slave, but a slave can recursively become master for other activities), *Heap-organized* (no master, work is distributed through data structures), *Pipeline* (work and intermediate results are passed in a predefined order), and *Systolic Multidimensional Pipeline* (same as Pipeline, but communication and synchronization is systolic; such implementations are usually two-dimensional).

4.6.4   Visualization using PIE

Although PIE can be used for both visualizing performance data and visualizing programming constructs, this section will focus on the first. Starting from the visualization of programming constructs, the user of PIE selects objects to be observed during execution by a mouse-click. PIE presents performance data in a variety of ways. Simple views include histograms and timeline formats. However, PIE's principal format for visualizing performance data is a Gantt Chart with time on the horizontal axis and individual processes ordered on the vertical axis. Each line (process) contains colored rectangles indicating what the process is currently doing. The names of these actions are taken from the corresponding programming constructs. If this information is not available, the rectangles are colored black (working) or white (switched off).

In a different view, *processor utilization*, a Gantt Chart contains processors instead of processes. In this way PIE visualizes which processor is executing which execution thread for a single computation.

4.6.5   Future Plans

According to announcements of its authors, PIE will be improved in many areas. Future plans include:

1. performance improvement;

2. various new sensor types (hardware and hybrid sensors), to reduce perturbation (probe effect);

3. design and implementation of a performance degradation prevention advisor and procedures for automatic techniques for performance degradation avoidance;

4. improvement of the interactive run time facilities;

5. additional options for customizing monitoring by users; and

6. support of different programming languages (currently only Pascal is supported).

4.7 TOPSYS

TOPSYS (Tools for Parallel Systems – Baker et al. [27]) is an integrated tool environment for increasing productivity when using and programming parallel computers.

4.7.1 Design Goals

The TOPSYS project has three major goals:

1. The first goal is to simplify the usage and the programming of parallel systems for widening the range of applications for these types of machines. Massively parallel computers should become as easy to use as today's conventional (sequential) workstations.

2. The TOPSYS environment is intended to be portable, flexible and adaptable with respect to processor architectures, communication networks, programming languages, operating systems, and to support tools, monitoring techniques and abstraction levels.

3. The complexity of the tool environment implementation is to be decreased by an intensive coupling with other subsystems of the target parallel computer without influencing the functionality and performance of these subsystems.

4.7.2 TOPSYS Structure

TOPSYS consists of *tools, monitors* and a *kernel.* The basic structure of TOPSYS is a hierarchical layered model. This hierarchical model stems mainly from the idea of using all tools together with all monitoring techniques regardless of the specific implementation of their corresponding partners.

1. The top of the hierarchy is the *graphical user-interface*, namely the X window system, which is used by all tools.

2. The *tools layer* consists of the available tools. Currently the most important tools are:

> DETOP (Debugging Tool for Parallel Systems);
> PATOP (Performance Analysing Tool for Parallel Processing);
> VISTOP (Visualizing Tool for Parallel Processors);
> SAMTOP (Specification and Mapping Tool for Parallel Systems).

All tools are used in on-line mode; therefore, all monitored data and their representations are available at run time.

3. The *monitor layer* contains hardware monitors as well as software monitors.

4. At the bottom of the hierarchy is the *Multiprocessor Multitasking Kernel* MMK, with an integrated dynamic load balancer.

4.7.3   Interface Definitions

The integration idea has led to the definition of two important interfaces:

Monitor Interface is a command driven interface, which is used by applications of the tool layer to get run time information of processor nodes. All monitor types offer the same functionality to the upper layers and are therefore replaceable with each other. This feature allows a flexible adaptation of the TOPSYS tool environment to all application requirements with respect to program retardation. In addition, the measurement of the program slowdown during the use of different monitors allows the qualification and quantification of the appropriateness of different instrumentation techniques.

Tool Interface is a procedural interface. From this interface all tools can request run time information about the dynamic behavior of the parallel execution at the abstract level of the source program.

The transformation between the abstraction level of the monitor interface (machine code) and the abstraction level of the tool interface (source code) is done within a central layer of the tool environment. This layer is responsible for the management of specified events, actions and symbol table objects. It consists of functions that are used by all tools and functions specially implemented for one of the tools.

4.7.4   Visualization Facilities

Within the TOPSYS environment those tools which can be used for visualization of parallel programs are PATOP, for performance analysis, and VISTOP, for algorithm animation.

Different levels of abstraction offer the choice of different levels of detail granularity. TOPSYS users can look at the parallel program at the following levels:

System level: Performance measurements of all processor nodes are summed up to give a total system performance.

Node level: Performance measurements for each node are possible.

Task level: Performance data of all MMK objects of a parallel program (task, semaphore, etc.) is available.

Source code level: The performance of one sequential task can be observed, and the execution of several procedures of statements can be measured.

At each level of abstraction, PATOP can measure different performance values. Depending on the selected level these performance values are CPU (task), communication, synchronization or memory oriented. The performance data values of each measurement are represented graphically in two ways:

Histograms are used for time related diagrams. The user has the possibility to perform statistical operations on the displayed values.

Bargraphs display measurement values or results of statistical operations performed on measurement values in an easily comparable way.

With the specification tool SAMTOP the programmer is able to specify object interaction at the process level graphically. With VISTOP it is possible to validate and graphically understand the program execution. This tool is able to visualize and animate parallel programs at the level of object graphs containing MMK objects and operations. VISTOP, as an on-line animation system, collects the run time data via the TOPSYS tool interface using global breakpoints and state inspection calls, and graphically animates this data concurrently. Other available features of VISTOP are variable update frequency of the animation, free choice of animating objects forward or backward, and optional display of the states of the animated objects.

4.8 VIPS

VIPS (Visualization and Interactive Programming Support – Isoda et al. [65]) shows static and dynamic behavior of program execution, and especially visualizes complex data structures. VIPS is designed for Ada programs. The Ada compiler generates two intermediate language files: *Diana* and *Quadruple*. The VIPS preprocessor analyzes the Diana file for information about blocks and variables and passes this information, together with the Quadruple file, to the visualization subsystem. So, in addition to providing traditional debugging facilities, VIPS can graphically represent several views of Ada program execution in seven windows:

Data Window is the most complex and sophisticated window. Every time a procedure or function is invoked, one or more subwindows are created. There are three kinds of subwindows:

Block: Every time a subprogram is invoked, a subwindow containing declared variables is created. These subprograms are displayed in a calling stack that illustrates their relationships.

Temporary: Variable accesses not displayed in block subwindows are shown in temporary subwindows.

List: Dynamic data is displayed in list subwindows.

Graphical data can be represented either in standard or in user-defined figures. The standard figure is a rectangle for scalar variables and a list of rectangles for arrays and records. Within each record the name and the value of each variable is displayed, unless there are too many variables to display. In this case VIPS only shows the structure (of that part) of the aggregate variable, and the user can zoom into the structure on request. With the *Figure Description Language* (FDL) user-defined figures can be created that define how variables or a combination of variables should be represented.

Dataflow is shown as an interaction between graphical figures in the data window, where used variables are highlighted in two different ways to indicate either read or write access by an arrow pointing from the read to the written variable.

Figure-Definition Window displays a list of variables and the names of user-defined variables. Variables chosen to be displayed in the data window can also be specified in this window.

Editor Window is used for editing the source program.

Acceleration Window displays the current execution speed as a percentage of the fastest possible. The user can set the execution speed with a mouse-click.

Program-Text Window displays the source program; the statement that was just executed is displayed in reverse video.

Block-Structure Window shows the nesting relationship of subprograms.

Interaction Window displays the interaction between the user and the program.

4.8.1   Improved Version of VIPS

The current version of VIPS (Shimomura, Isoda [112]) based on the experience of the first version, and emphasizes the visualization of linked lists. The biggest drawback of the first version was that it displayed data only in one way (each node's data values). Thus only a part of the whole list could be displayed. The current version not only solves this problem, by providing multiple views, but is also improved in several other ways:

Easy shape recognition: It is easy to recognize an erroneous shape caused by an erroneous pointer value of the linked list.

Easy change detection: Not only a change in the shape of the linked list, but also changes in the value of data elements of each node can be recognized with ease.

Selective Display: Linked lists that are large and of several node types are difficult to understand. The current version of VIPS is able to display a linked list in three different ways:

- The *whole list* displays the overall structure of all nodes, while hiding node data. Each node is represented by a small box which is highlighted when any element of it is changed.

- The *selective list* displays only a part of the whole list, guaranteeing faster graphical update times.

- The *partial list* is recommended only for a small number of nodes because it displays each node's data in a way similar to VIPS first implementation.

The user can switch between these display types by selecting new sublists.

Rapid Drawing: Fast response time is essential for efficient debugging, and has therefore been dramatically improved.

Language support: While the first version of VIPS supported only the debugging of Ada programs, the current version uses UNIX's symbolic debugger, DBX, to execute the program and to obtain the required information. Therefore, common languages, like Fortran, C, and C++ can be debugged with VIPS.

4.9 SMILI

Most visualization systems are only usable for a relatively small level of parallelism: implementations are limited to 8 or 16 nodes because the user cannot extract much information if he has to scan a 128 point Kiviat graph or a Gantt Chart, which displays processor activity of 128 processors. Therefore SMILI (Scientific visualization in Multicomputing for Interpretation of Large amounts of Information – Khanna, McMillin [66]) uses cartoon faces, also known as *Chernoff Faces* (Chernoff [40]), to visualize a multi-dimensional data space.

The facial expression is controlled by several parameters, such as the shape of the face, the size of the eyes or the tie, and the form of the mouth.

A sample of points in p-dimensional space is represented by a collection of faces. The ability to relate faces to emotional reactions carries a mnemonic advantage. Although faces can be abused by portraying wrong information, a proper assignment may actually enhance the analysis procedure. For instance, a smile of the mouth can be used to represent a success/failure variable or the length of the nose can be a synonym for the execution time.

SMILI, which is based on the X Window System, is a post-mortem visualization system — there must be a trace file containing the relevant performance data. Before the system can be used for a particular program analysis, the parameters of the faces need to be tuned to the special needs of the user. This is accomplished by starting with a random parameter choice, and then varying one parameter at a time, while keeping the others fixed. After working with the system for some time, the user learns to relate facial expressions to program states. Therefore, in subsequent experiments, performance comparison is just pattern recognition.

4.10 Summary

A summary of many important visualization products can be found in the following table, which provides information about the aspects described in Chapter 2.

1. The first column gives the name of the visualization system or tool.

2. The second column contains references to the bibliography.

3. The third column marks systems especially designed for parallel computers with a 'P'. Systems without a 'P' might also be used for parallel systems, but are restricted there in their functionality.

4. The fourth column indicates the main purposes of the system: animation of the algorithm, visualization of program behavior, visualization of performance aspects, program creation, or program debugging.

5. The fifth column distinguishes between systems performing visualization at the run time of the investigated program (real-time visualization), and systems that use a trace file for post-mortem visualization.

6. The last column indicates possible types of visualization: The listed systems can visualize either the algorithm underlying the investigated program, the interconnection topology, the process mapping, data of the program (variables, input and output data), the program code, the program's behavior, or various performance aspects. Many systems are able to visualize multiple aspects.

System	Reference(s)	Target[5]	Purpose[6]	Time[7]	Kind[8]
BALSA	Brown, Sedgewick [38]		A	RM	AC
BALSA-II	Brown [34]		A	R	AC
BUGNET	Wittie [122]	P	D	R	B
BELVEDERE	Hough, Cuny [61]	P	AD	M	B
COMET	Kumar [8]	P	P	M	P
CLAM/CLAMSHELL	Foulser, Gropp [48]		A	M	D
DPM	Miller [84]	P	P	RM	P
FAUST	Guarna et al. [54]		PC	M	CP
IPS/IPS-2	Miller, Yang [86], Miller et al. [85]	P	P	M	MP
IVE	LaPolla [70, 71]	P	B	R	B
IVT/DT	Gupta et al. [20]		B	R	D
MAKBILAN	Rubin et al. [110]	P	AD	R	MP
MAP/SHMAP	Brewer et al. [32], Dongarra et al. [44]	P	B	M	A
MATRIX	Paul, Poplawski [97]	P	P	M	D
PARAGRAPH	Heath, Etheridge [57]	P	BP	M	TBP
PARET	Nichols, Edmark [95]	P	PC	R	P
PASTIS	Müller et al. [91]		A	R	D
PAW/Q+	Melamed, Morris [83], Funka-Lea et al. [52]	P	PC	R	B
PAWS	Pease et al. [98]	P	PC	R	M
PECAN	Reiss [103]		AC	M	C
PEGASYS	Moriconi, Hare [90]		C	R	C
PIE	Rudolph, Segall [111], Lehr et al. [75]	P	P	RM	B
POKER	Snyder [113]	P	CD	R	TD
PPUTTS	Fowler et al. [49]	P	PD	M	B
PREFACE	Bernstein et al. [29]	P	P	RM	P
PROVIDE	Moher [87]		D	R	DB
PV	Brown et al. [124]		BC	R	DC
SLALOM	Rover et al. [107]	P	P	R	P
SPD/SPM	Po et al. [99]		P	R	P
SMILI	Khanna, McMillin [66]		BP	M	P
TANGO	Stasko [115]	P	AP	M	AP
TASKGRAPHER	El-Rewini, Lewis [47]	P	P	M	TP
TOPSYS	Baker et al. [27]	P	PCD	R	DP
TRACEVIEW	Malony et al. [80]	P	P	M	P
VIPS	Isoda et al. [65], Shimomura, Isoda [112]		BD	R	D
VISTA	Tuchman et al. [118]	P	D	RM	D
VOYEUR	Socha et al. [114]	P	AD	M	AB
ZEUS	Brown [35]		A	R	A

[5]Target: System was (mainly) designed for Parallel Machines.

[6]Purpose: Algorithm Animation, Program Behavior or Performance or Creation, Debugging.

[7]Time: Real-Time or Post-Mortem Visualization.

[8]Kind: Algorithm, Topology, Mapping, Program Data or Code or Behavior, Performance.

Chapter 5

PARAGRAPH

The visualization system PARAGRAPH (Heath, Etheridge [57]) will be used extensively in the second and third part of this book for demonstrating various types of visualizing parallel scientific programs: parallel IDeC methods for the solution of ordinary differential equations and parallel integration methods, respectively. For this reason this chapter contains a detailed explanation of PARAGRAPH's display facilities along with an overview of the available options. Displays are grouped into five main categories: utilization, communication, tasks, miscellaneous and application-specific. Each group will be described separately.

The examples of this chapter are primarily concerned with the numerical solving of initial value problems of ordinary differential equations, which will be the main topic of the second part of this book.

5.1 Execution Control

Before the execution of the program to be visualized the user selects one or more of the available displays described below. To begin the execution of the program the user presses the Start button. The animation then proceeds to the end of the trace file unless the user interrupts the animation by pressing the Pause/Resume button. Pressing this button again continues the animation. A Step button is available to single-step through the trace file (one event per step). The Start button can be used again to restart the animation from the beginning, thus providing the user with the opportunity to execute the program as often as desired. Finally a particular time interval can be specified, and the visualization can be optionally stopped each time a specified event occurs.

5.2   Utilization Displays

Utilization describes how effectively the processors are used and how evenly the computational work is distributed among the processors. Each processor always has one of the following *states*, which are mapped to the specified colors, using a traffic light metaphor:

Idle (red = stop) is the state of a processor that has suspended execution while waiting for a message that has not yet arrived, or that has finished execution at the end of the run.

Overhead (yellow = caution) is the state of a processor that is executing some communication task but not waiting for a message.

Busy (green = go) is the state of a processor that is executing some (computational) part of the program without communication (i.e. doing "useful" work).

There are seven displays that deal with processor utilization:

Utilization Count shows the total number of processors in each of the three states listed above. The numbers of the processors are on the vertical axis; time is on the horizontal axis (Figure 5.1).

Utilization Gantt Chart provides the state information as a function of time for each processor individually (Figure 5.2).

Utilization Summary displays the percentage of time over the entire length of a run that each processor spent in the three states (Figure 5.3).

Utilization Meter provides the same information as the utilization count display, but rather than scrolling with time, it changes dynamically in place (Figure 5.4).

Concurrency Profile informs the user about the percentage of time (vertical axis) a certain number of processors (horizontal axis) were in a given state (one state per display, Figure 5.5).

Kiviat Diagram gives a geometric depiction of the utilization of individual processors and the overall load balance across processors (Figure 5.6).

Streak Display shows processors on the horizontal axis and winning/losing streaks on the vertical axis. Busy is always considered winning, while idle is always considered losing. Overhead may be classified as either winning or losing. As the current streak for each processor grows, the corresponding vertical bar rises (winning streak) or falls (loosing streak)

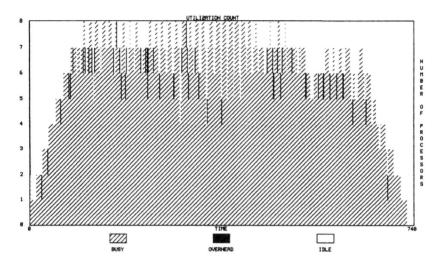

FIGURE 5.1: Utilization Count Display

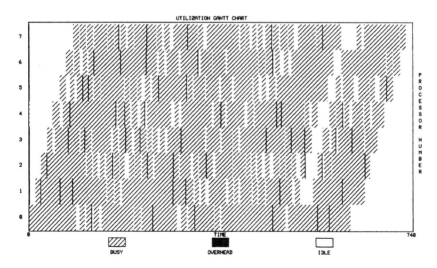

FIGURE 5.2: Utilization Gantt Chart

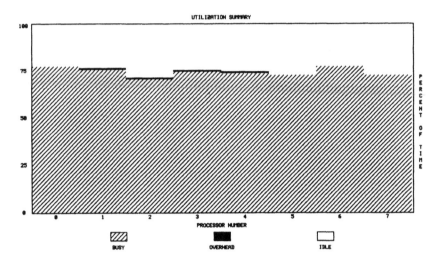

FIGURE 5.3: Utilization Summary

FIGURE 5.4: Utilization Meter

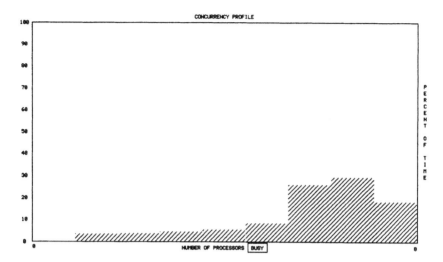

FIGURE 5.5: Concurrency Profile

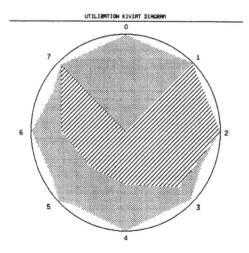

FIGURE 5.6: Kiviat Diagram

from the horizontal axis. When a streak for a processor ends, its bar returns to the horizontal axis to begin a new streak. At the end of the run, the longest streaks (both winning and losing) at any point during the run are shown for each processor (Figure 5.7).

5.3 Communication Displays

Interprocess communication displays are helpful in determining the frequency, volume, and overall pattern of communication:

Communication Traffic Display shows the total amount of communication (vertical axis, expressed either by message count or byte volume) as a function of time (horizontal axis). The curve is the total of all currently pending messages of all (optionally, one particular) processor(s) (Figure 5.8).

Space-Time Diagram depicts the interactions among processors (vertical axis) through time (horizontal axis). Processor activity is indicated by lines drawn solidly if the corresponding processor is busy, and by a blank if idle. Messages between processors are depicted by lines between the sending and the receiving node. The color of each line indicates message size, message type or distance traveled (Figure 5.9).

Message Queues Display shows the size of the message queue (expressed either in message count or byte size shown on the vertical axis) of each processor (horizontal axis) changing over time. A processor's input queue contains all messages sent to that processor that were not yet received. [1] The diagram displays both the current queue size (darker shade) and the maximum queue size (lighter shade) of each processor's message queue (Figure 5.10).

Communication Matrix Display represents messages in a two-dimensional array. The rows and columns correspond to the sending and receiving processor for each message. During the animation, a message is shown by the appropriate square being colored at the sending time and erased at the receiving time; as in the space-time diagram the chosen color indicates message size, message type or distance traveled. This display is also capable of showing the cumulative communication volume for the entire run (Figure 5.11).

Communication Meter serves a similar purpose as the utilization meter. It provides the same information as the communication traffic display but changes in place and therefore does not require too much screen space.

---

[1] A message is not treated "received" before the receiving node has asked for it.

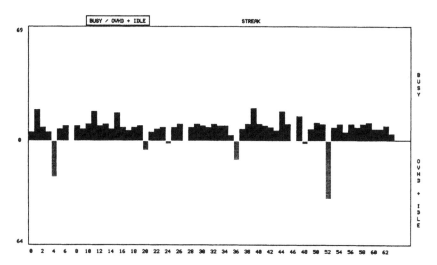

FIGURE 5.7: Streak Display

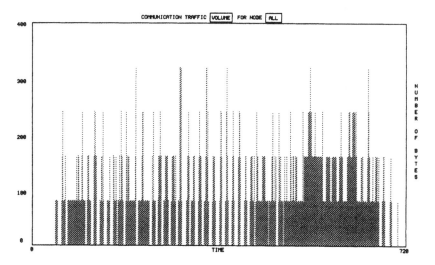

FIGURE 5.8: Communication Traffic Display

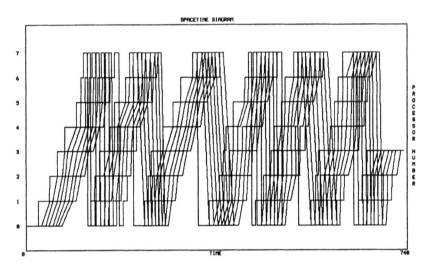

FIGURE 5.9: Space-Time Diagram

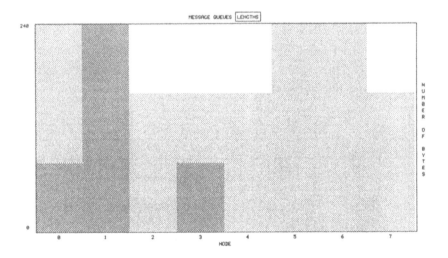

FIGURE 5.10: Message Queues Display

Animation Display represents the multiprocessor system by a graph whose nodes represent processors, and whose lines represent communication links. Colors indicate each node's status (busy, idle, sending, receiving). Lines between the source and destination nodes are drawn when a message is sent, and erased when it is received. After the run has finished, this display shows a summary of all communication links used during the execution (Figure 5.12).

Hypercube Display is similar to the animation display, but it additionally provides a number of layouts for the nodes in order to exhibit more clearly communication patterns corresponding to the various networks that can be embedded in a hypercube.[2]

Network Display depicts interprocess communication in terms of various network topologies. Unlike the previous two displays, it shows the actual path that each message takes, which may include routing through intermediate nodes. Obviously, depicting message routing through a network requires knowledge of the interconnection topology, which the user has to provide.

Node Statistics Display is a detailed graphical report for a single, user-selected processor. Time is on the horizontal axis and the chosen statistic (source and destination code, message type, message length, Hamming distance) is on the vertical axis.

Color Code Display permits the user to select the type of statistic that is displayed in the space-time diagram and the communication matrix display. The choices include the message type, the size of the message in bytes, and the distance between the source and destination nodes. Apart from the Hamming distance, which is suitable for hypercubes, PARAGRAPH offers distance functions for various mesh, torus, and tree topologies.

5.4 Task Displays

In order to use the *task displays* the user has to define "tasks" within the program by using special PICL routines to mark the beginning and ending of each task and assigning it a self-chosen number, i.e. the user is free to choose what is meant by a task. Four different task displays are available:

---

[2]This does not require that the interconnection network of the machine on which the program is executed actually be a hypercube, it is just intended to highlight the hypercube structure as a matter of potential interest.

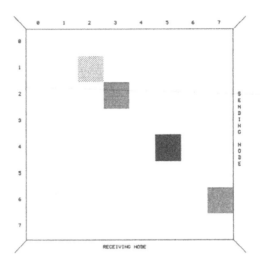

FIGURE 5.11: Communication Matrix Display

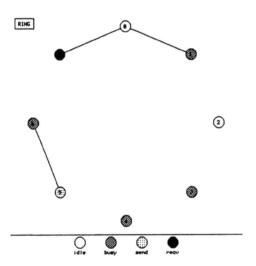

FIGURE 5.12: Animation Display

Task Count Display shows during the animation the number of processors (vertical axis) that are executing a given task (color coded on the horizontal axis). At the end of the run, the display shows the average number of processors that were executing each task over its lifetime (Figure 5.13).

Task Gantt Chart depicts the task activity of individual processors (vertical axis) by a horizontal bar chart, in which the color of each bar indicates each processor's currently running task as a function of time (horizontal axis) (Figure 5.14).

Task Status Display represents tasks in a two-dimensional array of squares ordered in rows. This two-dimensional layout may be used to save screen space for other displays. At the beginning all squares are white; as each task begins, its corresponding square is lightly shaded; after its completion the corresponding square is darkly shaded (Figure 5.15).

Task Summary, defined only at the end of a run, indicates the duration of each task (horizontal axis) as a percentage of the overall execution time (vertical axis). The bar of each task ranges from its earliest beginning to its latest completion (by any processor) (Figure 5.16).

5.5   Other Displays

Several additional displays either do not fit into any or actually fit into more than one of the three categories described so far:

Phase Portrait depicts the relationship between communication (vertical axis) and processor use (horizontal axis) over time. At any given point in time, the percentage of processors in the busy state and the percentage of the maximum volume of communication together define a single point in the two-dimensional plane. This point changes with time as communication and processor utilization vary, thereby plotting a "trace" in the plane (Figure 5.17).

Critical Path Display, similar to the space-time diagram, highlights the longest serial thread in the parallel computation.

Processor Status Display gathers information about processor utilization, communication, and tasks into a single display. In each of the four subdisplays the processors are represented by a two-dimensional array of squares ordered in rows. The subdisplays contain the current state, the task being executed, the message volume being sent, and the message volume awaiting receipt for each processor (Figure 5.18).

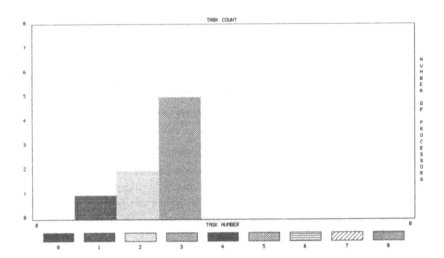

FIGURE 5.13: Task Count Display

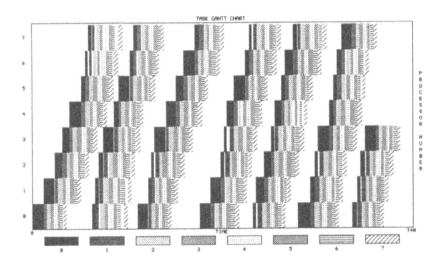

FIGURE 5.14: Task Gantt Chart

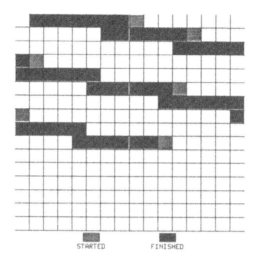

FIGURE 5.15: Task Status Display

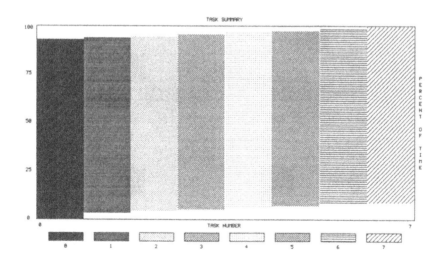

FIGURE 5.16: Task Summary Display

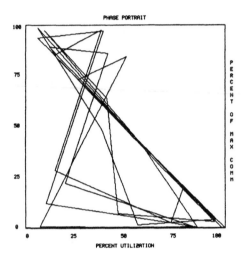

FIGURE 5.17: Phase Portrait Display

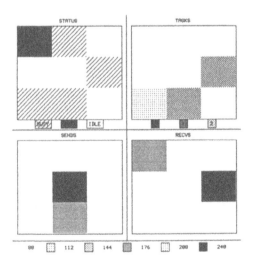

FIGURE 5.18: Processor Status Display

Digital Clock provides the elapsed time both as a numerical value and as a colored horizontal bar.

Trace Display may not only be useful for visualization, but also for debugging and single-stepping. It textually prints each event read from the trace file.

Statistical Summary Display gives exact numerical values for processor utilization and communication, for both individual processors and aggregates of processors.

Coordinate Information Display contains information produced by mouse clicks on the other displays. Many of the previously mentioned displays respond to mouse clicks by printing in this display the coordinates (in units meaningful to the user) of the point at which the cursor is located at the time the button is pressed.

5.6 Application-specific Displays

PARAGRAPH is extensible in that users can add application specific displays of their own design. To achieve this, PARAGRAPH contains calls at appropriate points to routines that provide initialization, data input, event handling, drawing, etc. for an application specific display. Of course, writing these necessary routines requires knowledge of programming X Window applications.

For the visualization of two scientific programs – implementations of parallel IDeC methods and parallel integration algorithms – new application-specific displays were developed. Their description can be found in chapter 10 and 16 respectively.

5.7 Options

The execution behavior and visual appearance of PARAGRAPH can be customized by the user in a number of ways:

1. Processors can be arranged in natural, Gray code, or user-defined order. Individual names can be given to each processor.

2. Windows, representing time along the horizontal axis can scroll smoothly or jump scroll by a user-specified amount.

3. The relationship between simulation time and the timestamps of the trace events can be chosen.

4. The trace file can be started and/or stopped at user-defined points.

5. A smoothing factor for Kiviat Diagrams and Phase Portrait displays can be selected.

6. A *slow motion slider* is provided that enables the user to control visualization speed.

7. All displays are laid out for both color displays and monochrome displays (using patterns instead of colors). When colors are used, the color palette may be chosen by the user.

8. All the chosen parameters and settings can be saved to an environment file, which is automatically loaded when PARAGRAPH is started.

Part II

Visualization of
Parallel IDeC Methods

This part of the book describes the authors' experience concerning the visualization of Iterated Defect Correction (IDeC) methods. Iterated Defect Correction is, among other applications, a special technique for accelerating the convergence and thus iteratively improving the accuracy of procedures for the numerical solution of initial value problems (IVPs) of systems of ordinary differential equations (ODEs). Chapter 6 contains an introduction to parallel IDeC methods.

The investigated IDeC methods differ in their kind of parallelization (Sections 6.5, 6.6, 6.7) and in characteristics influenced by the respective target machines (Chapter 8). Each parallel IDeC program run produces a trace file (Chapter 9) which is subsequently used by PARAGRAPH (Chapter 5, Chapter 10) as an input for post-mortem visualizations. PARAGRAPH was chosen because it offers a great variety of visualization types. Moreover, it can be extended by new types of views appropriately adapted to the particular needs of parallel IDeC methods (Chapter 10).

Specific examples of the visualization of different parallel IDeC methods as well as an evaluation of those special techniques and their results can be found in Chapter 11 and 12.

Chapter 6

Parallel IDeC Methods

6.1 Introduction

Compared to parallel methods in the field of numerical linear algebra, there exist only few parallel procedures for the practical solution of initial value problems (IVPs) of systems of ordinary differential equations (ODEs). This fact can be explained primarily by the *inherently sequential* nature of any forward step method for the numerical solution of such problems. Massive parallelism in this field either requires a priori knowledge of a special structure of the respective problem or a great number of redundant calculations. In this way the potential power of a parallel computer system is normally utilized only to a modest extent. Most existing methods are efficient only for a low degree of parallelism (fully utilizing only 2–5 processor elements).

Iterated Defect Correction (IDeC) is a special technique for accelerating the convergence and thus iteratively improving the accuracy of procedures for the numerical solution of IVPs of ODEs. IDeC schemes are characterized, among other properties, by their flexibility regarding possible fields of application, high efficiency and advantageous stability properties.[1] Detailed descriptions, analyses and discussions of IDeC procedures can be found in Auzinger et al. [136], Frank et al. [148, 149, 150], Frank, Ueberhuber [156, 157, 158], Ueberhuber [188].

Several published methods for the parallel solution of ODE IVPs have unfavorable stability properties which require the choice of impractically small step sizes. Such methods suffer from a significant reduction of their potential performance gain. IDeC methods do not have this handicap. When they are based on appropriate implicit discretization schemes, they boast very favorable stability and convergence properties. In particular the property of B-convergence (introduced by Frank et al. [151, 152, 153, 154, 155]) makes IDeC methods par-

---

[1]Procedures of this type can be used to solve stiff as well as non-stiff initial and boundary value problems of ordinary or partial differential equations.

ticularly promising candidates for the main component of efficient solvers for
non-linear stiff ODEs (Frank et al. [148], Auzinger et al. [136]).

Some procedures from the class of IDeC methods for solving IVPs of stiff
differential equations have been demonstrated to be at least as efficient as
widely used, very efficient software for this task, such as, for instance, the
BDF-Codes by Hindmarsh (Ueberhuber [188]).

Parallelization of Defect Corrections as proposed by Augustyn, Ueberhuber
[134] leads to numerous parallel IDeC procedures. They possess various quali-
ties which make them better or less suited for different target computers.

In this book a meta-algorithm is presented which leads to many promising
parallel IDeC procedures. Emphasis lies on the following points:

- General *algorithmic* description of the IDeC scheme (emphasizing features
  relevant for parallelization) in the form of Fortran 90 meta-algorithms.

- Description of an appropriate parallelization technique.

In this way the foundation for the development of specific parallel IDeC methods for the numerical solution of ODE IVPs is laid.

6.2   The IDeC Procedure

6.2.1   The Defect Correction Principle

The *Iterated Defect Correction* (IDeC) principle is a general acceleration technique applicable to the numerical solution of a large class of operator equations
$Fz = 0$, where $F : E \to E^0$ is a nonlinear operator and E and E^0 are normed
linear spaces (see, for instance, Frank, Ueberhuber [158]). It will be discussed
in the following as a basis for the construction of efficient algorithms for the
numerical solution of initial value problems (IVPs) of systems of ordinary differential equations (ODEs).

A given ODE IVP (called the *original problem*)

$$y' = f(t, y), \quad t \in [t_0, t_N]$$
$$y(0) = y_0, \quad\quad y', y, f \in \mathbb{R}^m \tag{6.1}$$

is solved on the grid

$$\mathbf{G} := \{t_\nu : t = t_\nu h, \nu = 0, 1, \ldots, N\}; \quad h := (T_N - T_0)/N$$

in an iterative way by the following steps:

1. Numerical solution of the IVP (6.1) with a typically "simple" (low effort)
   basic method suitable for solving systems of ODEs, such as, for instance,
   an explicit or implicit Runge-Kutta-method.

2. Estimation of the defect (the residual with respect to (6.1)) of the approximate solution.

3. The defect calculated in Step 2 is used for the construction of a new IVP, the so-called *neighboring problem*, whose exact solution is known explicitly.

4. Numerical solution of the neighboring problem despite the knowledge of its exact solution. The method used for this calculation is either the basic method used already in Step 1 or some other "simple" (cheap) method.

5. The difference between the exact and the numerical solution of the neighboring problem (i.e. the accurately known discretization error of this numerical solution) serves as an *estimate* for the unknown error of the basic solution of the original problem.

6. The error estimate obtained in Step 5 is used for the *correction* (improvement of the accuracy) of the basic solution.

7. Steps 2 to 6 are repeated iteratively as long as an increase of accuracy can be achieved (or a user-specified error tolerance is met).

There are theoretical and practical limits for the iterative improvement of an approximate numerical solution. After a certain number of defect-correcting steps, it is not possible to achieve any further improvement of accuracy (Frank, Ueberhuber [158]). These limitations will be discussed in detail later on.

The rough outline given above shows that the IDeC procedure is not just *one* special method for solving ODE IVPs but potentially yields a whole class of methods. Depending on which basic method is chosen and which method is chosen for solving the neighboring problems, special instances from this class are obtained. Still, these are not the only possibilities for variation, as will be shown later. The enormous flexibility of the defect correction principle makes possible the construction of a variety of algorithms well suited for particular problem classes or as the backbone of efficient general IVP solvers.

Figure 6.1 illustrates the defect correction principle by showing the results[2] of one special IDeC method applied to the simple test equation $y' = -y$.

6.2.2 IDeC Meta–Algorithm

The following description of the Iterated Defect Correction scheme will be given in single steps from which the whole procedure can be composed. Each step

---

[2]In Figure 6.1, results of one particular IDeC process are shown on an undistorted scale. However, some parameters of the procedure have been chosen in a way which makes the effects of the underlying defect corrections more clearly visible.

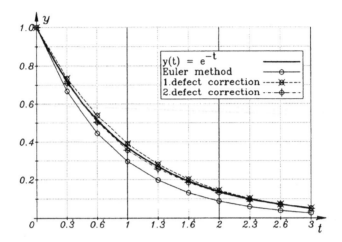

FIGURE 6.1: Solution of $y' = -y$, $y(0) = 1$ with an IDeC Procedure

is explained using a combination of mathematical and algorithmic means, as the aim of this book is to lay the foundation for IDeC implementations. Mathematical descriptions are used only as far as they are needed to understand details of IDeC implementations. Mathematical analyses can be found in some other publications (Frank, Ueberhuber [156, 157, 158], Frank et al. [148, 149], Auzinger et al. [136], Ueberhuber [188]).

In the following a pseudo-code in Fortran 90 style, which has been devised to be as self-explanatory as possible, is used for the algorithmic description of IDeC procedures for solving ODE IVPs.

Basic Solution: The initial value problem (6.1) is solved with a basic method $\mathcal{M}^b$ on a grid $\mathbf{G} := \{t_\nu : t_\nu = \nu h, \ \nu = 0(1)N; \ h = (t_N - t_0)/N\}$. The values of the approximate solution obtained with $\mathcal{M}^b$ on this grid are indicated by $\zeta^0 = (\zeta_0^0, \ldots, \zeta_N^0)$, $\zeta_\nu^0 \in \mathbb{R}^m$. The following pseudo-code illustrates the process:

```
h        = (tn - t0)/n
t_start = t0
t_end   = t0 + h
y_start = y0
grid(0) = t_start
zeta(0) % t(0) = y_start

DO i = 1, n
   zeta(0) % t(i) = basic_method (f,t_start,t_end,y_start)
   grid(i) = t_end
```

```
    t_start = t_end
    t_end   = t_start + h
    y_start = zeta(0) % t(i)
END DO
```

with the following correspondences: $\text{tn} \ldots t_N$
$\quad\quad\quad\quad\quad\quad\quad\quad\quad\quad\quad\quad\quad\quad\quad\quad\quad\quad \text{t0} \ldots t_0$
$\quad\quad\quad\quad\quad\quad\quad\quad\quad\quad\quad\quad\quad\quad\quad\quad\quad\quad \text{n} \ldots N$
$\quad\quad\quad\quad\quad\quad\quad\quad\quad\quad\quad\quad\quad\quad\quad\quad\quad\quad \text{y0} \ldots y_0$
$\quad\quad\quad\quad\quad\quad\quad\quad\quad\quad\quad\quad\quad\quad\quad\quad\quad\quad \text{f} \ldots f.$

The variable `zeta` is of the user-defined type `grid_value`

```
TYPE grid_value
   TYPE (function_value) :: t(0:n)
END TYPE grid_value
...
TYPE (grid_value) :: zeta(n_DeCs)
...
```

The type `function_value` is specified in the following way

```
TYPE function_value
   REAL, DIMENSION (m)  ::  y
END TYPE function_value
```

For the description of the method it is not relevant, however, whether a scalar differential equation or a system of equations is dealt with.

Defect Estimation: The defect (residual) of the discrete approximate solution $\zeta^0 \in \mathbb{R}^{m \times (N+1)}$ with respect to the given ODE $y' = f(t, y)$ cannot be calculated immediately. First of all the discrete points $\zeta_0^0, \ldots, \zeta_N^0$ representing the numerical solution have to be projected onto a differentiable function $\nabla \zeta^0(t)$. Only now can the *defect* d^0, i.e. the residual of $\nabla \zeta^0$ with respect to the given ODE, be obtained by substituting $\nabla \zeta^0$ for y in the differential equation (6.1):

$$d^0(t) := (\nabla \zeta^0)'(t) - f(t, (\nabla \zeta^0)(t)), \quad t \in [t_0, t_N]. \tag{6.2}$$

The projection ∇ can be defined, for example, via piecewise polynomial interpolation of degree M. This will be the method of choice for the following examinations (cf. Section "Choice of the Interpolation Scheme", p. 76).

The following piece of code shows the steps needed to calculate the defect at the grid points of $\mathbb{G}$:

```
TYPE (grid_value) :: p, dp, defect
...
dp % t = derivative_of_interpolation (zeta(0),t)
defect % t = dp % t - f(t(0:n), zeta(0) % t)
```

Components of a structure are referenced in Fortran 90 style as

$$structure\ \%\ \mathtt{t}$$

which can be read as "*structure* (evaluated) at $\mathtt{t}$".

Error Estimation: The defect d^0 obtained in the previous step can now be used to calculate an estimate of the global error $\zeta^0 - \Delta y$ of the approximate solution ζ^0. Δy denotes the discrete values of the continuous solution y of the original problem (6.1) at the points of grid $\mathbf{G}$.

First, a so-called *neighboring problem*[3] is constructed by means of the defect (6.2), which has the interpolating function $\nabla \zeta^0$ as its exact solution:

$$\begin{aligned} y' &= f(t,y) + d^0(t), \quad t \in [t_0, t_N] \\ y(t_0) &= y_0. \end{aligned} \tag{6.3}$$

Then both the newly defined initial value problem (6.3) *and* the original problem (6.1) are solved with some method $\mathcal{M}$ (denoted as $\mathtt{method\_2}$ in the pseudo-code) on the same grid $\mathbf{G}$ as used for the basic method $\mathcal{M}^b$. The approximate solution of the neighboring problem obtained in this way is called π^0, while the approximate solution of the original problem produced by $\mathcal{M}$ is called ξ. A standard choice would be $\mathcal{M} = \mathcal{M}^b$, in which case $\xi = \zeta^0$ is already available (however, we will continue the general case $\mathcal{M} not = \mathcal{M}^b$).

```
TYPE (grid_value) :: xi, pi(0:n_DeCs-1)
...
xi % t(0)    = y_start
pi(0) % t(0) = y_start

DO i = 1, n
   y_start_xi = xi % t(i-1)
   y_start_pi = pi(0) % t(i-1)
   xi % t(i)    = method_2 (f,grid(i-1),grid(i),y_start_xi)
   pi(0) % t(i) = method_2 (f_neighbor_problem,grid(i-1),   &
                            grid(i),y_start_pi)
END DO
```

---

[3]It is called *neighboring problem* because the exact solutions of the original problem and the neighboring problem are close together (provided the basic method for solving (6.1) does not produce a totally useless approximation for y).

The function subprogram `f_neighbor_problem` is specified by:

```
FUNCTION f_neighbor_problem (t,y) RESULT (f_t_y)

    !      uses global variable zeta,
    !            global function f

    REAL,                   INTENT (IN) :: t
    TYPE (function_value),  INTENT (IN) :: y
    TYPE (function_value)               :: f_t_y

    TYPE (function_value)               :: p, dp

    p     = interpolate (zeta(0),t)
    dp    = derivative_of_interpolation (zeta(0),t)
    f_t_y = f(t,y) + (dp - f(t,p))

END FUNCTION f_neighbor_problem
```

The global error $\pi^0 - \zeta^0$ of the approximate solution π^0 of the neighboring problem is known *without* the otherwise unavoidable uncertainty, because the exact solution of the neighboring problem (6.3), i.e. $\nabla \zeta^0$, is known. $\pi^0 - \zeta^0$ is normally a good approximation of the *unknown* global error $\xi - \Delta y$ of the approximate solution ξ of the original problem, as both solutions are close together. A theoretical explanation of this fact is given by Frank, Ueberhuber [158].

Correction: The error estimate $\pi^0 - \zeta^0$ of the previous step is used to improve the accuracy of the approximate solution ξ.

In the following identity (where Δy denotes the discrete values of the continuous solution y of the original IVP (6.1) at the points of grid **G**)

$$\Delta y = \xi - (\xi - \Delta y)$$

the unknown global error $\xi - \Delta y$ is substituted by the known global error $\pi^0 - \zeta^0$ of the neighboring problem. In this way an improved solution can be obtained:

$$\zeta^1 := \xi - (\pi^0 - \zeta^0) = \zeta^0 - (\pi^0 - \xi).$$

In algorithmic description:[4]

```
zeta(1) % t = zeta(0) % t - (pi(0) % t - xi % t)
```

----

[4]Note that here a Fortran 90 feature has been used which allows operations on entire arrays. `zeta(1) % t` is an array consisting of elements of type `function_value`. This is the case for the other terms as well. When converting into an executable program, note that addition and subtraction have to be defined (overloading the operators + and −) for the type `function_value`!

Iteration: By constructing a further neighboring problem in the same way as above but using the improved solution ζ^1, calculating the respective defect $d^1(t)$ and subsequently an estimate for the global error of ζ^1, it is possible to improve ζ^1 in the same manner as the accuracy of ζ^0 has been increased in the previous step:[5]

$$\zeta^2 := \xi - (\pi^1 - \zeta^1) = \zeta^1 - (\pi^1 - \xi).$$

This process can be continued iteratively, resulting in further improvements in the accuracy:

$$\zeta^{j+1} := \xi - (\pi^j - \zeta^j) = \zeta^j - (\pi^j - \xi).$$

If $\mathcal{M}^b$ is a method of order p and $\mathcal{M}$ a method of order q with $q \leq p$, then under certain assumptions[6] (Frank, Ueberhuber [158]) the numerical solution ζ^j is of the order $\min\{p + jq, J\}$ where the maximum achievable order J is mainly determined by properties of the interpolatory function used for defining the defect of ζ^j. For example, for piecewise polynomial interpolation at equidistant points, J is equal to the degree of the polynomials. In any case, there is an upper limit for the continuation of the iterative correction process which should generally not be exceeded.

IDeC Meta – Algorithm

Assembling the separate steps described above leads to a general meta-algorithm:

```
TYPE function_value
    REAL :: y  ! one scalar equation is assumed (to simplify matters)
END TYPE function_value


TYPE grid_value
    TYPE (function_value) :: t(0:n)
END TYPE grid_value


FUNCTION IDeC_step (f,t0,tn,y_start,n,n_DeCs) RESULT (IDeC_iterate)
    INTEGER,                INTENT (IN) :: n_DeCs  ! number of DeC steps
```

---

[5]In view of the fact that ζ^1 is usually more accurate than ζ^0, the solution of the neighboring problem based on ζ^1 is also nearer to the exact solution of the original problem (6.1). Thus $\pi^1 - \zeta^1$ allows a better estimation of the global error of ξ and gives the opportunity of improving ζ^1.

[6]The order results quoted in this section apply to non-stiff ODEs. In the case of stiff ODEs it is much more complicated to obtain comparable results. Consequently less rigorous characteristics are available so far for IDeC methods applied to stiff problems (Auzinger et al. [136]).

```
   INTEGER,                 INTENT (IN) :: n         ! n + 1 gridpoints
   REAL,                    INTENT (IN) :: t0, tn  ! interval
   TYPE (function_value),   INTENT (IN) :: y_start
   TYPE (function_value)                 :: IDeC_iterate

   INTERFACE
      FUNCTION f(t,y) RESULT (right_hand_side)
      REAL,                    INTENT (IN) :: t
      TYPE (function_value),   INTENT (IN) :: y
      TYPE (function_value)                 :: right_hand_side
   END INTERFACE

   INTEGER              :: i, j
   REAL                 :: h
   REAL                 :: grid(0:n)
   TYPE (grid_value)    :: xi
   TYPE (grid_value)    :: zeta(0:n_DeCs), pi(0:n_DeCs-1)

   h = (tn - t0)/n
   grid(0) = t0
   xi % t(0)              = y_start
   pi(0:n_DeCs-1) % t(0) = y_start
   zeta(0:n_DeCs) % t(0) = y_start

   DO i = 1, n
      grid(i) = grid(i-1) + h
      zeta(0) % t(i) = basic_method (f,grid(i-1),grid(i),zeta(0)%t(i-1))
      xi % t(i) = method_2 (f,grid(i-1),grid(i),xi%t(i-1))
   END DO

   DO j = 0, n_DeCs - 1
      DO i = 1, n
         pi(j) % t(i) = method_2 (f_neighbor_problem,grid(i-1),grid(i), &
                           pi(j) % t(i-1))
      END DO
      zeta(j+1) % t = zeta(j) % t - (pi(j) % t - xi % t)
   END DO

   IDeC_iterate = zeta(n_DeCs) % t(n)

CONTAINS

   FUNCTION f_neighbor_problem (t,y) RESULT (f_t_y)

      !     uses global variable zeta (defined in   IDeC_step),
      !           global variable j   (defined in   IDeC_step),
      !           global function f   (parameter of IDeC_step)

      REAL,                    INTENT (IN) :: t
      TYPE (function_value),   INTENT (IN) :: y
      TYPE (function_value)                 :: f_t_y
```

```
TYPE (function_value)                    :: p, dp

p    = interpolate (zeta(j),t)
dp   = derivative_of_interpolation (zeta(j),t)
f_t_y = f(t,y) + (dp - f(t,p))

END FUNCTION f_neighbor_problem

END FUNCTION IDeC_step
```

Choice of the Interpolation Scheme

In the previous definition of ∇ it was assumed that, as a part of every correction step, the respective ζ^j is interpolated over the *whole* interval $[t_0, t_N]$. At first sight this assumption seems to be a serious drawback with respect to the solution of initial value problems. In most cases it is algorithmically extremely disadvantageous to apply just *one* global method of interpolation to data over the whole interval $[t_0, t_N]$. The number of interpolatory points needed in such a procedure would be unacceptably high, not to mention the fact that the IDeC procedure would lose a lot of its flexibility. To prevent this, the interval $[t_0, t_N]$ is actually subdivided into a number of subintervals

$$[T_0, T_1], [T_1, T_2], \ldots, [T_{K-1}, T_K] \tag{6.4}$$

and the IDeC principle is applied separately in a forward step manner to an appropriately defined IVP over each of these intervals. These intervals do not have to be of an a priori determined length; they can be determined at run time. Thus, the prerequisites for an automatic *adaptation of the step size* are given.

To allow the separate consecutive processing of the IVP (in the sense of a forward step algorithm) over the intervals (6.4), only *local interpolation schemes* are eligible, i.e. interpolation methods where $\nabla\zeta^j(t)$ can be calculated from values $\{\zeta_\nu^j\}$ inside a certain vicinity of t. The most natural choice is piecewise polynomial interpolation on (6.4). However, if the operator ∇ is defined in this way then $\nabla\zeta^j(t)$ is only a piecewise differentiable function over $[T_0, T_K]$, i.e. $(\nabla\zeta^j)'(t)$ has jump discontinuities at the points $T_1, \ldots T_{K-1}$ where the polynomial segments join. The neighboring problems (6.3) are therefore IVPs with piecewise defined right-hand sides. This creates some difficulties for the analysis of such IDeC methods (Frank, Ueberhuber [158]) but does not impair their practical applicability (provided $\mathcal{M}$ is a single step method, which means that no information from backward grid points is used).

The points where the segments of the piecewise polynomial of degree M join define a subgrid $\overline{\mathbf{G}}$ of $\mathbf{G}$ (provided $N = K \cdot M$)

$$\overline{\mathbf{G}} := \{T_i : T_i = T_0 + iH, i = 0(1)K; \ H := Mh\}.$$

The polynomial pieces are defined on the intervals

$$I_i := [T_{i-1}, T_i], \quad i = 1(1)K$$

by the requirement

$$P_i^j(t_\nu) = \zeta_\nu^j, \quad \nu = (i-1)M, \ldots, iM.$$

Consequently the intervals I_i are called *interpolation intervals*.

As was pointed out by Frank, Ueberhuber [158], it is essentially the choice of the interpolation operator ∇ which determines the maximum attainable order of an IDeC method. If ∇ is defined by Lagrangian interpolatory polynomials, the maximum attainable order is given by the degree M of the polynomials. The asymptotic argument which leads to this result remains valid if a step size control is applied to adapt the grid $\overline{\mathbf{G}}$ to the pecularities of a specific initial value problem. If the subgrid on each interval I_i consists of $M+1$ equidistant points, then the grid $\mathbf{G}$ so defined may be considered as one element of an infinite grid sequence (despite the fact that the step lengths $H_i := T_i - T_{i-1}$ will be chosen a posteriori during the course of the computation) and consequently the theoretical results apply. If, however, the number of grid points is not the same in different interpolation intervals or the gridpoints on any interval I_i are not equidistant, no theoretical results exist to predict the asymptotic behavior of the discretization errors.

Forward Step IDeC Methods

First, all calculations are performed on the initial interval, i.e. for the IVP

$$\begin{aligned} y' &= f(t, y), \quad t \in [T_0, T_1] \\ y(T_0) &= y_0. \end{aligned} \tag{6.5}$$

The quantities $\zeta_0^0, \ldots, \zeta_M^0$ and $\xi_0, \ldots, \xi_M$ are obtained by applying $\mathcal{M}^b$ and $\mathcal{M}$ to problem (6.5). The numerical solutions $\pi_0^0, \ldots, \pi_M^0; \pi_0^1, \ldots, \pi_M^1; \ldots; \pi_0^{j_{max}}, \ldots, \pi_M^{j_{max}}$ of the neighboring problems

$$\begin{aligned} y' &= f(t, y) + (P_1^j)'(t) - f(t, P_1^j(t)), \quad t \in [T_0, T_1] \\ y(T_0) &= y_0 \end{aligned} \tag{6.6}$$

$(j = 0, 1, 2, \ldots, j_{max})$ are used to calculate successively the quantities

$$\zeta_0^1, \ldots, \zeta_M^1; \zeta_0^2, \ldots, \zeta_M^2; \ldots; \zeta_0^{j_{max}}, \ldots, \zeta_M^{j_{max}}$$

as explained above.

Assume that both $\mathcal{M}^b$ and $\mathcal{M}$ are one step methods with increment functions Φ^b and Φ. This assumption will make the following discussion more transparent. On the second interval $[T_1, T_2]$ the approximations $\zeta^0_{M+1}, \ldots, \zeta^0_{2M}$ and $\xi_{M+1}, \ldots, \xi_{2M}$ are thus obtained by

$$
\begin{aligned}
\zeta^0_{M+1} &= s^b_1 + h \cdot \Phi^b(t_M, s^b_1; h, f) \\
\zeta^0_{\nu+1} &= \zeta^0_\nu + h \cdot \Phi^b(t_\nu, \zeta^0_\nu; h, f), \quad \nu = M+1\,(1)\,2M-1
\end{aligned} \tag{6.7}
$$

and

$$
\begin{aligned}
\xi_{M+1} &= s_1 + h \cdot \Phi(t_M, s_1; h, f) \\
\xi_{\nu+1} &= \xi_\nu + h \cdot \Phi(t_\nu, \xi_\nu; h, f), \quad \nu = M+1\,(1)\,2M-1
\end{aligned} \tag{6.8}
$$

respectively. The numerical solutions of the corresponding neighboring problems are given by

$$
\begin{aligned}
\pi^j_{M+1} &= s^j_1 + h \cdot \Phi(t_M, s^j_1; h, f + d^j) \\
\pi^j_{\nu+1} &= \pi^j_\nu + h \cdot \Phi(t_\nu, \pi^j_\nu; h, f + d^j), \quad \nu = M+1\,(1)\,2M-1
\end{aligned} \tag{6.9}
$$

$(j = 0, 1, 2, \ldots, j_{max})$. The starting values $s^b_1, s_1, s^0_1, s^1_1, \ldots$ are obtained from the computations on the first interval $[T_0, T_1]$ (note that the subscript of $s^b_1, s_1, s^0_1, \ldots$ corresponds to the subscript of the gridpoint T_1)

$$
\begin{aligned}
s^b_1 &:= \zeta^0_M \\
s_1 &:= \xi_M \\
s^j_1 &:= \pi^j_M, \quad j = 0, 1, 2, \ldots, j_{max}.
\end{aligned} \tag{6.10}
$$

In the same way the process is continued over $[T_2, T_3], [T_3, T_4], \ldots, [T_{K-1}, T_K]$. Thus, the IDeC procedure has been reformulated in a forward step manner. This formulation suggests another type of IDeC scheme: it seems to be advantageous to use only the most accurate approximation at T_1 (i.e. $\zeta^{j_{max}}_M$) as the starting value in (6.7), (6.8), and (6.9):

$$
\begin{aligned}
s^b_1 &:= \zeta^{j_{max}}_M \\
s_1 &:= \zeta^{j_{max}}_M \\
s^j_1 &:= \zeta^{j_{max}}_M, \quad j = 0, 1, 2, \ldots, j_{max}.
\end{aligned} \tag{6.11}
$$

In this case the IDeC scheme may be interpreted as a special one step method on the grid $\{T_0, T_1, \ldots, T_K\}$, i.e., all manipulations involved in the defect correction process on the interval $[T_{i-1}, T_i]$ may be considered as one single step of a one step method, the last and most accurate defect correction iterate at the gridpoint T_i being defined as the result of this step.

    For obvious reasons the procedure of performing defect correction with starting values (6.10) is called *global connection strategy*, whereas the choice of the initial values (6.11) leads to IDeC methods with *local connection strategy*.

Global Connection Strategy

In any case the IDeC process is first applied to the IVP (6.5) on the subinterval $[T_0, T_1]$ exactly as described above. Then, not only $\zeta_M^{j_{max}}$ is saved, but also ξ_M and ζ_M^j, π_M^j, $j = 0(1)j_{max}$. These values are needed as initial values for the respective IVPs (original problem and neighboring problems) on the second subinterval $[T_1, T_2]$. In the meta–algorithm listed above this means substituting the initialization

```
xi % t(0)             = y_start
pi(0:n_DeCs-1) % t(0) = y_start
zeta(0:n_DeCs) % t(0) = y_start
```

by

```
xi % t(0)             = xi_of_last_interval % t(m)
pi(0:n_DeCs-1) % t(0) = pi_of_last_interval(0:n_DeCs-1) % t(m)
zeta(0:n_DeCs) % t(0) = zeta_of_last_interval(0:n_DeCs) % t(m)
```

when integrating the IVPs on the second and all following interpolation intervals. This results in an IDeC procedure where in each step the whole interval $[t_0, t_N]$ is not interpolated globally by one smooth function, but piecewise with different functions. Figure 6.2 illustrates the outcome of such a procedure.[7]

Advantages of the global connection strategy are:

- Global error estimates are available as a spin-off without additional calculations.

- It is possible to parallelize the procedure. This aspect will be discussed in detail in Chapter 6.3.

- There are practically no restrictions concerning the methods $\mathcal{M}$ and $\mathcal{M}^b$. For example, even multistep methods could be used.

- Failure of the IDeC principle in a few subintervals of the system of differential equations, i.e. defect corrections leading to worse instead of better estimates, does not necessarily lead to failure over the whole interval.

- For stiff problems, global connection is usually to be preferred (see Auzinger et al. [136] for an explanation).

---

[7]In Figure 6.2 the ratios between the approximate solutions $\zeta^0, \zeta^1, \zeta^2, \ldots$ are shown on a distorted scale to make them more clearly visible.

Disadvantages:

- The number of defect correction iterations has to be determined *before* the procedure is started and cannot be increased at run time with a tolerable amount of additional calculations. In this case the complete IDeC procedure would have to be recalculated if not all intermediate values were kept in storage (which is normally not feasible in forward step methods for ODE IVPs) and all missing values are to be determined additionally.

- Defect correction iterates can be used only for estimates of the *global discretization error*, yet step size adaptation using *global error estimates* is economically unfeasible (Shampine, Watts [181]).

An adaptive step size selection mechanism for IDeC methods has to be based on additionally calculated estimates of the local error of the basic method $\mathcal{M}^b$. Local errors of the iteratively improved solutions $\zeta^1, \zeta^2, \dots$ cannot be utilized for step size selection purposes in an efficient way.

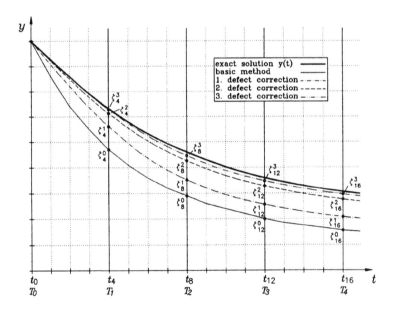

FIGURE 6.2: IDeC Procedure with Global Connection Strategy ($M = 4$)

IDeC Meta–Algorithm with Global Connection Strategy

```
FUNCTION IDeC (f,t0,tn,y_start,m,n_DeCs) RESULT (IDeC_iterate)

   INTEGER,                  INTENT (IN) :: n_DeCs  ! number of IDeC steps
   INTEGER,                  INTENT (IN) :: m       ! m + 1 gridpoints
   REAL,                     INTENT (IN) :: t0, tn  ! interval
   TYPE (function_value),    INTENT (IN) :: y_start
   TYPE (function_value)                 :: IDeC_iterate

   INTERFACE
      FUNCTION f(t,y) RESULT (right_hand_side)
      REAL,                  INTENT (IN) :: t
      TYPE (function_value), INTENT (IN) :: y
      TYPE (function_value)              :: right_hand_side
   END INTERFACE

   REAL                  :: h, t_start, t_end
   INTEGER               :: j
   TYPE (function_value) :: xi_last
   TYPE (function_value) :: pi_last(0:n_DeCs-1)
   TYPE (function_value) :: zeta(n_DeCs), zeta_last(0:n_DeCs)

   t_start = t0
   h = choose_first_step_size()
   xi_last = y_start
   pi_last(0:n_DeCs-1) = y_start
   zeta_last(0:n_DeCs) = y_start

   DO
      t_end = t_start + h
      CALL IDeC_step (f,t_start,t_end,pi_last,xi_last,   &
                      zeta_last,m,n_DeCs)
      IF ( test_step_size() ) THEN    ! step is accepted
         IF (t_end >= tn)     EXIT    ! terminate loop
         t_start = t_end
      END IF
      h = choose_step_size()
   END DO

   IDeC_iterate = zeta_last(n_DeCs)

CONTAINS

   SUBROUTINE IDeC_step (f,t_start,t_end,pi_start,xi_start,   &
                         zeta_start,m,n_DeCs)

      !     uses global variable zeta (defined in   IDeC),
      !           global variable j    (defined in   IDeC),
      !           global function f    (parameter of IDeC)

      INTEGER,                 INTENT (IN)    :: m, n_DeCs
```

```
REAL,                    INTENT (IN)    :: t_start, t_end
TYPE (function_value), INTENT (INOUT) :: xi_start
TYPE (function_value), INTENT (INOUT) :: pi_start(0:n_DeCs-1)
TYPE (function_value), INTENT (INOUT) :: zeta_start(0:n_DeCs)

INTEGER                             :: i
REAL                                :: h
REAL                                :: grid(0:m)
TYPE (grid_value)                   :: xi
TYPE (grid_value)                   :: pi(n_DeCs)

h = (t_end - t_start)/m
grid(0) = t_start
xi % t(0) = xi_start
pi(0:n_DeCs-1) % t(0) = pi_start(0:n_DeCs-1)
zeta(0:n_DeCs) % t(0) = zeta_start(0:n_DeCs)


DO i = 1, m
   grid(i) = grid(i-1) + h
   zeta(0) % t(i) = basic_method (f,grid(i-1),grid(i),zeta(0)%t(i-1))
   xi % t(i)      = method_2 (f,grid(i-1),grid(i),xi%t(i-1))
END DO

DO j = 0, n_DeCs - 1
   DO i = 1, m
      pi(j) % t(i) = method_2 (f_neighbor_problem,    &
                     grid(i-1),grid(i),pi(j)%t(i-1))
      zeta(j+1) % t(i) = zeta(j) % t(i) - (pi(j) % t(i) - xi % t(i))
   END DO
END DO

xi_start = xi % t(m)
pi_start(0:n_DeCs-1) = pi(0:n_DeCs-1) % t(m)
zeta_start(0:n_DeCs) = zeta(0:n_DeCs) % t(m)

END SUBROUTINE IDeC_step

FUNCTION f_neighbor_problem (t,y) RESULT (f_t_y)

   !    uses global variable zeta (defined in   IDeC),
   !         global variable j    (defined in   IDeC),
   !         global function f    (parameter of IDeC)

   REAL,                    INTENT (IN) :: t
   TYPE (function_value), INTENT (IN) :: y
   TYPE (function_value)              :: f_t_y

   TYPE (function_value)              :: p, dp

   p      = interpolate (zeta(j),t)
```

```
dp      = derivative_of_interpolation (zeta(j),t)
f_t_y = f(t,y) + (dp - f(t,p))

END FUNCTION f_neighbor_problem

END FUNCTION IDeC
```

Local Connection Strategy

Here, in contrast to the global connection strategy, the IDeC procedure is restarted at the beginning of each subinterval $[T_{i-1}, T_i]$, $i = 1(1)K$. The best approximate value of the previous interval, i.e. ζ_M^{jmax}, serves as the initial value. The outcome of such a procedure is illustrated in Figure 6.3.[8]

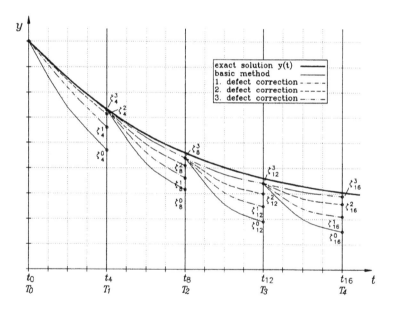

FIGURE 6.3: IDeC Procedure with Local Connection Strategy

Advantages of the local connection strategy are:

- The number of defect correction iterations can be rechosen in each subinterval $[T_{i-1}, T_i]$ (independent of the previous course of calculations). It is possible, for instance, to incorporate a mechanism (*order control*) in

---

[8]In Figure 6.3 the ratios between the approximate solutions $\zeta^0, \zeta^1, \zeta^2, \ldots$ are shown on a distorted scale to make them more clearly visible.

an IDeC algorithm which stops the iterative computation of $\zeta^0, \zeta^1, \ldots$ for each subinterval $[T_{i-1}, T_i]$ as soon as a given error tolerance is met. Thus an IDeC procedure with variable step size and variable order is obtained.

- If the calculations on each subinterval $[T_{i-1}, T_i]$ are based upon the best approximate solution on the previous subinterval, a better overall accuracy is to be expected (intuitively) than with the global connection strategy.

- An estimate of the local error of the best approximate value $\zeta_M^{j_{max}}$ is easily available (Ueberhuber [188]). However, when treating *stiff differential equations* it is necessary to take care not to *under*estimate local errors. This is because of superconvergence phenomena observable when IDeC methods are applied to stiff problems of a specific category.

Disadvantages:

- In contrast to the global connection strategy it is not possible to obtain a global error estimate.

- If the IDeC principle fails on account of a special difficulty in the differential equations, the error level can be considerably worse than with the global connection strategy, as practical experiments (Fasching [147]) have shown. This results from the fact that in such cases the *worst* approximate value of the previous interval is used as the initial value for the following subinterval.

- Local connection is usually inferior in accuracy in stiff situations (Auzinger et al. [136]).

- Only single step methods are suitable as procedures $\mathcal{M}^b$ and $\mathcal{M}$ for numerically solving the original problem and the neighboring problem.

- IDeC algorithms with local connection strategy *cannot be parallelized.* Thus they are not considered any further in this book.

A detailed description of IDeC algorithms with local connection strategy is not given in this book. They can easily be constructed by simply changing the initial vectors in any IDeC algorithm using the global connection strategy. An extensive discussion of the local connection strategy for the practical solution of stiff differential equations can be found in the IDeC literature (Ueberhuber [188]).

Possible Variations

There are a number of ways to construct special IDeC algorithms, i.e. instances of the IDeC meta-algorithms specified above. A few of these possibilities are listed in the following as examples, though the list is far from complete. Unfortunately not all of these variants possess theoretical foundations. Some of them have only been tested in experiments (Frank et al. [150]).

- Methods for solving the original problem $(\mathcal{M}^b)$ and the neighboring problems $(\mathcal{M})$ can be chosen according to the nature of the differential equations. Implicit methods will be used for stiff problems, explicit methods (for example Runge-Kutta schemes) for non-stiff problems. Depending on the connection strategy certain restrictions have to be observed. For the local connection strategy multi-step methods cannot be used, as the linking of the subintervals causes problems.

  Generally speaking, the order of $\mathcal{M}^b$ and $\mathcal{M}$ should be chosen rather low, otherwise

  - the required level of accuracy is mainly achieved not by the IDeC principle but by $\mathcal{M}^b$. An application of the IDeC principle (based on the global connection strategy) with a higher order method $\mathcal{M}^b$ may be useful in situations where only global error estimates for the results of $\mathcal{M}^b$ are to be calculated.

  - order control for IDeC algorithms based on the local connection strategy is only possible in a coarse mode as the order sequence of the iterates is given by

    $$p, p + q, p + 2q, \ldots, \min\{p + jq, J\}$$

    with J being the maximum attainable order.

  - the overall effort and time needed for such an IDeC method will be too high to be competitive with other methods.

  - the degree of parallelism becomes too small to be of interest.

- In each iteration of a defect correction process a different method for the solution of the neighboring problem could be used. Furthermore it would be possible to use different methods in every subinterval $[T_{i-1}, T_i]$ allowing the local adaptation of an IDeC algorithm to peculiarities of the differential equations by using methods (for instance stiff/non-stiff integrators) which are *locally* adequate.

- The interpolation scheme needed for the construction of the neighboring problems provides a large number of possible variations. In principle,

different procedures could be used for every iteration and in each interpolation interval. Furthermore the gridpoints of an interpolation interval
do not necessarily have to be equidistant. When the gridpoints inside the
intervals $[T_{i-1}, T_i]$ are distributed in a specific way (for instance, according
to a Radau node distribution) the following observation can be utilized:
in some specific situations (special types of stiff differential equations) it
turns out the sequence $\{\zeta^j\}$ converges to a fixed point ζ^* which coincides
with the solution of some collocation scheme which is known for its favorable properties in connection with stiff systems (Frank, Ueberhuber
[156], Ueberhuber [188]).

- It is possible to leave some of the gridpoints out of the interpolation mechanism. Such a procedure can be useful, for instance, if the discretization
  error of the solution methods shows an oscillatory behavior. For example,
  it would be possible to calculate $2M$ ζ-values in one subinterval, but to
  use only every second point for interpolation purposes in order to keep
  the error level as low as possible.

6.2.3   An IDeC Method for Parallelization

The following instance from the class of IDeC procedures has been chosen to
demonstrate the inherent possibilities of the IDeC principle for parallelization:

- The *global connection strategy* has been chosen as the only option which
  allows parallelization. With respect to the overall accuracy this constraint
  does not mean a serious drawback. Practical experiments have shown
  that the results of IDeC methods based on the local connection strategy
  are only slightly more accurate than those of IDeC methods based on the
  global connection strategy, and may even be worse (Fasching [147]).

- *Polynomial interpolation on equidistant grids* has been chosen because of
  its simplicity. The evaluation of interpolatory polynomials and their first
  derivatives on equidistant grids is conceptually easy and requires only a
  minor computational overhead (needed for multiplications with weight
  matrices). The same kind of of polynomial interpolation is used in each
  subinterval and for every defect correction.

  This seems to be a plausible choice, because all requirements of the theoretical examination of the respective IDeC procedure are satisfied (especially those concerning the asymptotic error behavior for $h \to 0$).

- The methods $\mathcal{M}^b$ and $\mathcal{M}$ are defined to be identical. In this case ξ does
  not need to be calculated separately, as $\xi = \zeta^0$.

- The *explicit Euler method* is used in each subinterval and for every iterate
  of the defect correction as the basic method $\mathcal{M}^b$ and the method $\mathcal{M}$, as

the highest degree of parallelization can be achieved with methods of order 1 (this fact will be discussed later on). This choice restricts the applicability of the resulting IDeC procedures to non-stiff ODEs.

It is sufficient to use explicit methods $\mathcal{M}^b$ and $\mathcal{M}$ to demonstrate the parallelization techniques for IDeC procedures. The following discussion is easily transferred to implicit methods.

6.3   Parallelization of IDeC Methods

6.3.1   Categories of Parallelism

Parallelism in methods for the solution of initial value problems of ordinary differential equations can be subdivided primarily into three categories (Gear [159, 161]):

Parallel methods: The algorithm for solving differential equations is built from separate statements and steps which can be executed in parallel to some extent. The level of abstraction of this form of parallelism is high.

The advantage of parallel methods is their independence of the degree of parallelism of the size of the system of differential equations to be solved. Even scalar differential equations can be solved on parallel computers with such methods. On the other hand only modest degrees of parallelization are attainable in this way, depending greatly on the method.[9] This is a result of the fact that the solution of initial value problems is an inherently sequential problem.

Parallelism in the system of differential equations: Depending on the number of available processors, the system of differential equations is divided into subsystems which allow parallel calculation. In contrast to parallel methods, massive parallelism can be obtained here for very large systems of differential equations (for instance PDE IVPs discretized using the method of lines approach). The degree of parallelism depends on the number of differential equations.

If the system of differential equations can be decoupled into independent subsystems, the resulting subsystems can be solved in parallel without the otherwise unavoidable communication delays, possibly with different methods.

Parallelization in this way is normally not possible for systems of stiff differential equations, as their solution requires implicit procedures. At

---

[9]If a high redundancy of calculations is accepted, high degrees of parallelism can be reached, but the time saved with respect to a comparable sequential procedure will be relatively small.

every step of a parallel implicit procedure every single processor has to communicate with all other processors (Gear [159]). However, all available parallel linear algebra software and tools can be used in the stiff case.

Parallelism in time: Parallelization is carried out in such a way that several subintervals can be treated in parallel. Among others, block methods, parallel predictor-corrector methods and "stabbing" methods can be subsumed under this type of parallelism (Gear [161]). There are no clear-cut boundaries to the other types of parallelism (for example, parallel predictor-corrector methods can be counted in the first category as well).

It is difficult to obtain high degrees of parallelism by this principle, unless considerable additional information on the problem is available or an extreme degree of inefficiency due to many redundant calculations is accepted. Procedures of this type have been introduced by Nievergelt [177] and Gear [161].

Mixed types of parallelism, especially of the first two categories, are of course possible.

6.3.2   Principles of Parallelization

Parallel methods for ODE IVPs can be derived on different levels introducing a "hierarchy of parallelism". The methods of the first three levels integrate the unmodified system

$$
\begin{aligned}
y' &= f(t,y), \quad t \in [t_0, t_N] \\
y(t_0) &= y_0
\end{aligned}
\tag{6.12}
$$

and yield a sequence of numerical approximations $\eta_i \approx y(t_i)$ on a set of grid-points $\{t_0, \ldots, t_N\}$. The fourth level methods subdivide the system (6.12) and integrate the resulting subsystems on separate processors. The numerical integration of every subsystem may be carried out with an individually chosen integration formula and an appropriate sequence of step sizes.

Parallelization principles of lower levels can be utilized by higher level methods.

Practical experiences and evaluations of the results of the various parallelization techniques, i.e. parallel algorithms and/or parallel software for ODE IVPs, are reported very infrequently in literature (Augustyn et al. [128]). In many cases only *formulas* are presented and analyzed with respect to their speed-up. Such analyses can be used as a basis for the selection of potentially interesting methods. However, the actual behavior of a piece of software produced on the basis of one of these methods may be impaired by many factors such as reduced stability regions. A realistic assessment is usually only possible by an evaluation based on modelling of the respective algorithms and target computers (Augustyn, Ueberhuber [133]) and/or extensive numerical experiments on parallel computers.

Level 1: Parallelization of Serial Programs

This type of parallelization is applied to algorithms already implemented as conventional (serial) computer programs. Potential parallelism in these ODE codes is exploited to some degree by parallelizing compilers (Zima, Chapman [193]). In particular parallelism of DO-loops can be detected and transformed into parallel instructions. The evaluation of "long" arithmetic expressions can be parallelized by splitting them up into subexpressions to be evaluated on separate processors.

This kind of parallelization can be applied, in principle, to all sorts of ODE solvers. The extent of instructions executable in parallel depends on the particular code. Especially methods which rely heavily on matrix and vector operations (like, for instance, BDF codes for stiff ODEs) are promising candidates for automatic parallelization.

Level 1 type parallelization is quite effective on SIMD computers. The power of MIMD computers can be exploited only to a minor degree in this way.

Level 2: Parallelization of Serial Algorithms

Level 2 is characterized by a *manual* analysis and parallelization of serial ODE codes. Normally a higher degree of parallelism can be achieved in this way. Parts of the code whose input and output do not rely on each other can be transformed into groups of instructions executable in parallel.

A bottleneck preventing higher degrees of parallelism is the step size selection mechanism needed in forward step ODE solvers. This important part of any automatic code for the integration of ODE IVPs requires complete information about the current step. Thus synchronization of all concurrent processes is required. Gear [159] suggests determining the step size and – if required – the order of the steps k *and* $k+1$ from the information available at step $k-1$. A subsequent inspection and revision of the chosen parameters of step $k+1$ during step $k+2$ should be performed. If the error estimates do not satisfy the error tolerance, *two* steps have to be rejected. Consequently this kind of mechanism is not well suited for ODEs which require many variations of the step size. For such problems, reduced efficiency is unavoidable with this approach.

Methods parallelized in this way are applicable even in cases of single differential equations. However, the number of instructions which can be executed in parallel is usually not very large. For instance, extrapolation methods are suitable for parallel computation. The entries of the extrapolation table can be computed on different processors.

Every step of an implicit ODE solver requires the solution of a system of nonlinear algebraic equations. If some variant of Newton's method is used for this purpose the computation of the elements of the $m \times m$ jacobian f_y can distributed on $p \leq m^2$ processors. The solution of the respective linear systems may be accomplished by one of the available parallel algorithms in linear algebra (see, for example, Dongarra, Sorenson [146], Dongarra et al. [2], Axelsson, Polman [137], Hockney, Jesshope [164], Ortega, Voigt [178]).

With respect to parallel computation, *singly-implicit methods* are often considered to be favorable candidates for efficient implementations (see, for example, Burrage [140], Butcher [142]). In these methods the system of linear equations to be solved in every Newton step can be partitioned into independent subsystems. The order of accuracy of these methods, however, can break down dramatically for certain types of stiff ODEs. A detailed analysis of the convergence behavior of such Runge-Kutta methods applied to general nonlinear system of differential equations has been given by Montijano [176] on the basis of the concept of *B-convergence* (Frank et al. [151, 152, 153, 154, 155]). The results of Montijano (revealing disappointingly low orders of B-convergence) demonstrate the necessity of careful analyses and/or computer experiments which include not only theoretically derived speed-up numbers.

There are suggestions to solve every component of (6.12) (or groups of components) on different processors, i.e. a parallelization on a large *task level* (see, for instance, Cash [144]). Speed-up of such a procedure depends critically on a number of factors (size of the system, subdivision into subsystems, speed of communication between processors etc.). That means turnaround time is not necesserily smaller when using such parallel methods as compared with sequential methods.

Level 3: Modification of Serial Algorithms

The third level is characterized by modifications of serial algorithms and newly developed algorithms aiming at optimum utilization of parallel computers. On uniprocessor machines such methods are usually not competitive with "classical" methods; they show their strength primarily on parallel systems.

Explicit and implicit *block methods* allow the parallel computation of some of the intermediate quantities required in the integration scheme: Birta, Abou-Rabia [139], Abou-Rabia [126], Burrage [140, 141], Cash [143, 144], Chu, Hamilton [145], Le Gall, Mouney [173], van der Houwen, Sommeijer [189], Shampine, Watts [181], Sommeijer et al. [185], Tam [186, 187], Worland [192] etc.

The degree of parallelism in such methods rises with the number of blocks. Block schemes of different orders can be combined to provide order control mechanisms.

Block schemes are not only possible for one step methods but also for multi-step methods. Depending on the system (6.12) and on the accuracy requirement the efficiency of computer codes based on such modified predictor corrector schemes may be greatly impaired by deteriorated stability properties. That means that reformulated multi-step methods with reasonable speed-up factors do not necessarily lead to a reduced overall computation time.

Level 4: Decomposition of the Problem

The basic idea of the fourth level methods is to decompose the whole system (6.12) into subsystems of one or just a few equations which are treated separately on the processors of a MIMD type parallel computer. Step sizes, orders and other parameters of the integration methods can be adjusted to the properties of the respective *sub*system. A concurrent computation of the instructions required for the solution of each subsystem brings a further speed-up effect.

There are mainly two groups of methods which are based on a decomposition of the problem:

Waveform Relaxation Methods: See Gear [160], Gear, Juang [162], Juang [165], Juang, Gear [166], Lelarasmee et al. [174], Sand, Skelboe [179], Skeel [182], White et al. [191] etc.

Multirate Methods: See Gear, Wells [163], Sand, Skelboe [179], Skelboe [183], Skelboe, Andersen [184], Wells [190] etc.

Waveform Relaxation Methods

Waveform relaxation methods are based on an iterative scheme extended over the whole interval of integration $[t_0, t_N]$. As a part of this scheme, decoupled subsystems are solved concurrently. Waveform relaxation methods have been developed for circuit analysis (Lelarasmee, Ruehli, Sangiovanni-Vincentelli Lelarasmee et al. [174]). The system (6.12) is decomposed into

$$y_i' = f_i(t, y_i, p_i), \quad i = 1, \ldots, r$$
$$y_i(t_0) = y_{0i} \tag{6.13}$$

with the decoupling vectors $p_i = (y_1, \ldots, y_{i-1}, y_{i+1}, \ldots, y_n)^\top$. Treating the vectors p_i as known parameters allows the simultaneous solution of the r independent IVPs (6.13). The relaxation process starts with an initial approximation over the whole interval $[t_0, t_N]$ needed to initialize the decoupling vectors $p_1, \ldots, p_r$.

The solution of stiff systems (6.12) usually requires the solution of systems of n algebraic equations in every step of an implicit scheme. The application of waveform relaxation methods allows the reduction to r systems of algebraic equations of smaller dimension and correspondingly leads to a greatly reduced computational effort.

Multirate Methods

The basic idea of multirate methods is the decomposition of the system (6.12) with respect to different kinds of variation, in particular for stiff problems (with different time scales). The system (6.12) is partitioned, for instance, into two coupled subsystems

$$y_1' = f_1(t, y_1, y_2), \qquad (6.14)$$
$$y_2' = f_2(t, y_1, y_2) \qquad (6.15)$$

where $y_1 \in \mathbb{R}^l$ comprises the rapidly varying components of y and $y_2 \in \mathbb{R}^{m-l}$ the slowly varying components (Skelboe [183]). Multirate methods numerically integrate such partitioned systems with different step sizes adapted to the rapidity of variation of the respective solution components. The decoupled computations for the subsystems are synchronized in such way that the largest step size is an integral multiple of all smaller step sizes (for instance $H = qh$). All computations can be structured into compound steps which involve one step for the components with the largest step size H and the respective number of steps needed for the more rapidly varying components to proceed from T_i to $T_i + H$ (for instance q steps in the example given above).

There are two strategies to integrate the complete system (6.12):

1. The computation starts with q steps (with step size h) of the numerical integration of subsystem (6.14) using extrapolated values of the component vector y_2. Afterwards one step (with step size $H = qh$) is performed to integrate system (6.15) numerically.

2. The computation starts with one step (with step size H) of the numerical integration of subsystem (6.15) using extrapolated values of the component vector y_1. Afterwards q steps (with step size $h = H/q$) are performed to integrate system (6.14) numerically.

The generalization of these strategies to a larger number of subsystems of (6.12) is straightforward.

For problems (6.12) with a loose coupling between the subsystems, multirate formulas are more efficient even for uniprocessor machines than the respective conventional method which treats the system as a whole (Wells [190]). For parallel computers all extrapolation steps are performed simultaneously and subsequently the subsystems are integrated concurrently on different processors.

The theoretical foundations of multirate formulas are not as extensive as those for conventional serial algorithms (Skelboe [183], Skelboe, Andersen [184]).

6.3.3 Parallelization of the IDeC Procedure

In Augustyn, Ueberhuber [134], the authors describe a basic concept for parallelizing the Iterated Defect Correction procedure with global connection strategy. The resulting parallel procedure belongs to the category of parallel methods. The following table explains this idea for a procedure with three defect correction iterates:

time-step	parallel tasks
1	basic procedure on $[T_0, T_1]$
2	basic procedure on $[T_1, T_2]$, 1st defect correction on $[T_0, T_1]$
3	basic procedure on $[T_2, T_3]$, 1st defect correction on $[T_1, T_2]$, 2nd defect correction on $[T_0, T_1]$
4	basic procedure on $[T_3, T_4]$, 1st defect correction on $[T_2, T_3]$, 2nd defect correction on $[T_1, T_2]$, 3rd defect correction on $[T_0, T_1]$
5	basic procedure on $[T_4, T_5]$, 1st defect correction on $[T_3, T_4]$, 2nd defect correction on $[T_2, T_3]$, 3rd defect correction on $[T_1, T_2]$
$\vdots$	$\vdots$

This table can be illustrated by the corresponding Figure 6.4, which represents the course of an IDeC procedure with three defect correction iterates. The circled numbers are the "time-steps" of the table above. Sections with the same number can be processed in parallel.

Task Dependencies

Many tasks of a parallel IDeC procedure strongly *depend* on each other. The first defect correction iterate on a certain interval $[T_{i-1}, T_i]$ can only be carried out after the basic method has been applied to this interval and the first defect correction on the previous interval has terminated. Similar dependencies apply to all other sections of the whole process. All these dependencies are shown in the respective dependency graph of Figure 6.5. The rules for the interpretation of this graph are simple:[10]

---

[10]The dependency graph is very similar to Figure 6.4. The vertical lines can be identified with the separate steps of the defect correction. In addition the graph contains the dependencies of the individual values.

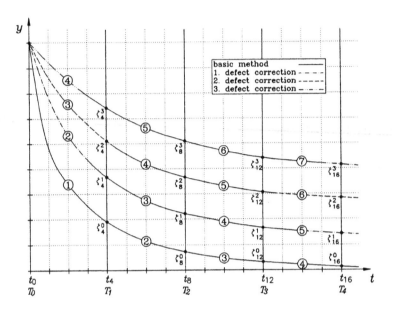

FIGURE 6.4: Parallelization Concept for IDeC Procedures

- Every node corresponds to an approximate solution calculated at a subinterval boundary T_i, $i = 0(1)K$.

  The intermediate values of the numerical solutions needed for the construction of the neighboring problems are *not* part of the graph. The inclusion of these subtasks would complicate the figure unnecessarily. A more detailed illustration of this type is not needed for the IDeC parallelization, as all calculations pertaining to one iterate on an interpolation interval can be seen as atomic units in this context.

- A directed edge between nodes i and j indicates that results of the task associated with the predecessor node i are needed for the task associated with the sucessor node j (see, for instance, Bertsekas, Tsitsiklis [138]).

The dependencies between the various tasks of an IDeC algorithm implicitly restrict the sequence of calculations. Execution of a task associated with one node of the dependency graph can only be started when all tasks associated with predecessor nodes have terminated and sent the required data. Nevertheless the designer of a parallel IDeC algorithm is free to choose between different variants of the sequence of calculations. This fact is essential for distributing tasks of an IDeC algorithm onto the processor elements of a parallel computer system.

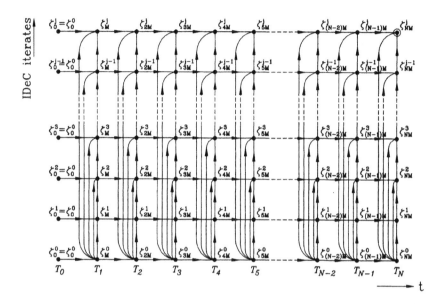

FIGURE 6.5: Dependency Graph for Parallel IDeC Methods with Global Connection Strategy

6.3.4   Step Size Control and its Effects

The definition of the numerical problem of finding an approximate solution of an IVP (6.1) has to include some criterion for distinguishing between approximations which are to be considered acceptable and those which are not. This is usually done by setting up a tolerance for the local or global error of the numerical result. The solution of such a problem (IVP *together* with some tolerance) requires the estimation of the errors being made during the integration and the adjustment of the step size (and other parameters of the method) to keep these errors within the given tolerance.

Estimates can be made for the local errors as well as for the global errors. Accordingly the control mechanisms can be based either on local or on global error estimates. Nearly all production codes for ODE IVPs compute only local error estimates and incorporate control devices based on these estimates.

The first question in connection with local error control is *which* local errors are to be estimated, for there is a multitude of local errors associated with IDeC algorithms.

When processing a subinterval I_i (either as a step of the basic method or an iterate of the defect correction) it makes sense to compute the local error of ζ^j $(0 \le j \le j_{max})$ only at the endpoints of the subintervals I_i, $i = 1(1)K$. The dependency graph in Figure 6.5 shows that step size control via the local error of ζ^j, $j > 0$ would severely restrict parallel processing. Every time a step size was rejected, not only the current results of the basic method would be wasted, but all previously computed defect correction iterates in that interval would have to be rejected and recomputed. Thus a change of step size would require a substantial computational overhead. Apart from that, the global connection strategy (essential for parallel IDeC methods) is not well suited for estimating the local errors of ζ^j, $j > 0$, because the calculation of each defect correction iterate starts with a different initial value in every interpolation interval. Thus, in an implementation of a parallel IDeC algorithm the step size control mechanism should be based on estimates of the local discretization error of the result ζ^0 of the basic method $\mathcal{M}^b$ alone.

The quality of the final result of an IDeC based ODE solver can be assessed using methods which do not rely on error estimates calculated for step size control purposes, such as, for instance, global error estimation (Shampine, Watts [181]). Global error estimates are a by-product of the global connection strategy. However, superconvergence effects of IDeC methods applied to stiff differential equations have to be taken into consideration. In this case the defect correction iterates may converge to a fixed point at a very high rate, which normally results in global error estimates being much too small. If this effect is not recognized and counteracted by the algorithm a serious deterioration of the reliability of an IDeC solver may be the consequence (Ueberhuber [188]).

The question how the rejection of a step affects the performance of a parallel IDeC algorithm can be answered easily by examining Figure 6.6:

- The bottom line corresponds to the basic method, and every line further up corresponds to a higher level of defect correction $j > 0$. The abscissa is *calculation time* and not the independent variable t of the system of differential equations.

- Each rectangle symbolizes all calculations associated with one interpolation interval and one level of defect correction j $(0 \le j \le j_{max})$.

- The temporal order of the rectangles results from the dependency graph in Figure 6.5. For example, ζ^1 can only be calculated for a subinterval if ζ^0 has already been computed for this subinterval and ζ^1 is available for the previous subinterval.

- Empty spaces in the direction of the time axis symbolize idle times.

Step size rejection results in a reiterated application of the basic method to those subintervals where the accuracy requirements cannot be met. This results

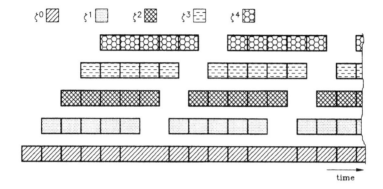

FIGURE 6.6: Effect of Step Rejections

in an increase of the overall run time of the basic method with respect to the subinterval causing trouble. Complying with the course of calculations indicated by the dependency graph, a rejected step of the basic method creates consecutive delays as illustrated in Figure 6.6. It turns out that a change of step size leads to a waiting period for *all* following defect corrections. Thus, extreme care must be taken that steps be rejected as seldom as possible – a great challenge for the designer of the step size control mechanism.

If only the basic method is subject to a step size control as in the previous assumption, it is possible to use conventional methods for this purpose. The classic "doubling-strategy" could be applied, i.e. local error estimation is made on the basis of extrapolation. This requires that the integration from T_{i-1} to T_i be done twice: as one step of length H_i and as two steps of length $H_i/2$. Such strategy leads to a considerable increase in the run time of the basic method. In practical implementations of parallel IDeC algorithms it is a must to parallelize all calculations needed for error estimation purposes (Gear [159]).

6.4 Load Distribution

The sequence of calculations of a parallel IDeC algorithm is fixed to a certain degree (Augustyn et al. [129]). The primary dependency graph of Figure 6.7 shows which quantities can be calculated in parallel at a given time step.[11]

Some freedom remains in the distribution of the tasks to be processed simultaneously on the nodes of a parallel computer. In principle, these tasks could be distributed arbitrarily to the available processors. Because tasks can be remapped at every time step, there are many possibilities for their (time

---

[11]Encircled numbers in the chart denote time steps.

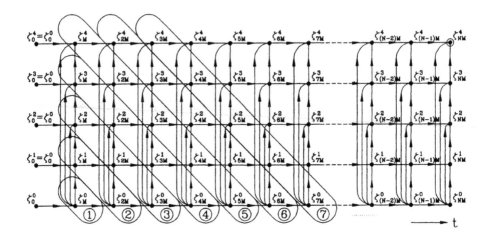

FIGURE 6.7: Temporal Order of Parallel Defect Corrections

dependent) distribution. In general, no one single method will be most efficient in all situations, but different methods or variants in different cases. Ideally, the parallel IDeC algorithm itself will choose dynamically the most efficient load distribution scheme (Krommer, Ueberhuber [172]) depending on parameters of the computational environment (computer architecture, number of available processors, communication delays between the processors, etc.). If the mapping and other algorithmic decisions are made in this way an *architecture adaptive algorithm* results (Augustyn et al. [129], Krommer, Ueberhuber [7]). Exact knowledge of the conditions which make one variant more effective than others is needed by the designer/developer of such an architecture adaptive parallel IDeC procedure.

In the following two different instances of the meta-algorithm described in Augustyn, Ueberhuber [130] are given. They all belong to the class of *parallel methods*, as opposed to *parallelism in the system of differential equations* or *parallelism in time* (Gear [159, 161], Nievergelt [177]). Each of these IDeC methods implements another load distribution scheme for the general method of parallel defect corrections (parallel IDeC method). For this purpose PF 90, an intuitive description language, derived from Fortran 90 will be used. It is specially designed for a simple, yet comprehensive description of parallel algorithms (Krommer, Ueberhuber [171]). It should be understood, that the PF 90 programs of the following sections are *not* runable programs or even production codes. They are only meant as formal descriptions of the respective IDeC methods.

6.5 IDeC Variant with Parallel Time Steps

IDeC methods with parallel time steps – referred to in the following chapters as IDeC$_{time}$ algorithms or methods – are based on the obvious load distribution scheme which allocates work units comprising all operations associated with one interpolation interval. Every processor is assigned the task of carrying out the basic method $\mathcal{M}^b$ as well as *all* calculations needed for the respective defect correction iterates on one specific interpolation interval.[12] Thus, every processor has to finish all calculations associated with one interpolation interval of the IDeC method before it can tackle a new task (i.e. before proceeding to an interpolation interval not yet treated).

The calculations of an IDeC$_{time}$ algorithm for p available processors and j_{max} defect correction iterations take their course according to the following load distribution scheme:

processor	calculation of
1	$\zeta^0, \ldots, \zeta^{j_{max}}$ on the intervals $[T_0, T_1], [T_p, T_{p+1}], \ldots$
2	$\zeta^0, \ldots, \zeta^{j_{max}}$ on the intervals $[T_1, T_2], [T_{p+1}, T_{p+2}], \ldots$
3	$\zeta^0, \ldots, \zeta^{j_{max}}$ on the intervals $[T_2, T_3], [T_{p+2}, T_{p+3}], \ldots$
$\vdots$	$\vdots$
p	$\zeta^0, \ldots, \zeta^{j_{max}}$ on the intervals $[T_{p-1}, T_p], [T_{2p-1}, T_{2p}], \ldots$

Synchronization of processes is straightforward in this IDeC-variant. Every instance of a process has to know the interpolation interval it is responsible for and the number of already computed defect correction iterates on this interval. A process calculating a new defect correction iterate can only be started as soon as the required initial value from the previous interpolation interval is available. As all processors are busy with calculations on distinct defect correction levels at any time[13], there will never be more than one approximate value of a defect correction level available as initial value for the following subinterval. In this way the processes are synchronized by the availability of initial values.

Figure 6.8 shows the start-up of an IDeC$_{time}$ method with four defect corrections in a time process diagram (Gantt Chart):

- Hatched *rectangles* represent basic steps and rectangles filled by other patterns represent defect correction steps on one interpolation interval.

---

[12]The length of an interpolation interval is determined by the processor responsible for it.

[13]That means it is not possible for more than one processor to do calculations at the same iterate level at any given moment of the whole run time.

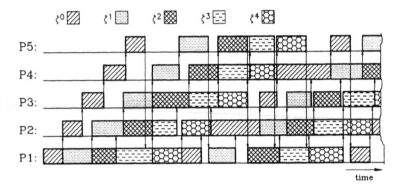

FIGURE 6.8: $IDeC_{time}$ – Time Process Diagram

- *Arrows* symbolize data (initial value) transmission between the processors. Transmitted are the approximate solutions at the right hand boundary of the interval just processed. Every processor is unable to continue its calculations on a certain level of defect corrections before it has received the initial values required for this level.

When they are ready to send, processors do not have to wait for a particular processor to be ready to receive the data. Data is broadcast (sent forth) "to whom it may concern".[14] After completing such a broadcast activity each transmitting processor can immediately resume its work.

6.5.1   Pros and Cons of $IDeC_{time}$ Methods

Advantages:

- Implementation is easy due to the simple synchronization mechanism.

- Only simple data structures are needed. For instance, *no queues* are required for an efficient implementation of $IDeC_{time}$ algorithms in contrast to the other variants – provided the communication system of the target machine permits the transmission of data without the necessity of waiting for a processor ready to accept it.

---

[14]This type of broadcast operation can be realized, for instance, by a queue of length 1 for every defect correction level. The transmitter puts the message into the queue and a receiver collects it from there. $IDeC_{time}$ methods guarantee that only one message is sent to a recipient processor responsible for a particular defect correction level at a time. For this particular transmission process queues with length 1 are sufficient.

- The smallest possible amount of data is to be communicated. Just one vector of initial values is needed for the calculations on every interpolation interval and every defect correction level. This is the only data to be transmitted in $IDeC_{time}$ algorithms.

- $IDeC_{time}$ algorithms adapt themselves automatically to the current number of available processors. This adaptation results from the fact that every processor works on a complete interpolation interval and every available processor is assigned one of these subintervals.

Disadvantages:

- When an integration step is rejected, the basic method has to be rerun one or several times. This repetition inevitably leads to an idle period on all other processors (the course of events is explained in detail in Augustyn et al. [129]). Other IDeC variants have opportunities to reduce these waiting periods.

- Statistical run time fluctuations lead in every case to a performance degradation as a result of the reduced processor utilization.

- It is not useful to provide more processors than defect correction levels plus one, as this would lead to idle processors due to the special synchronization mechanism. Massive parallelism is *not* possible with $IDeC_{time}$ algorithms.

On account of their disadvantageous properties, $IDeC_{time}$ methods do not utilize the potential capacity of a parallel computer to the same extent as other IDeC variants. $IDeC_{time}$ algorithms have higher total run times and a lower efficiency than $IDeC_{order}$ methods.

6.6 IDeC Variant with Parallel Defect Corrections

IDeC methods with parallel defect corrections – referred to in the following chapters as $IDeC_{order}$ methods – distribute the calculations needed for the individual defect correction iterates to the available processors. According to this load distribution principle each processor is assigned the calculation of the values of one *row* of the primary dependency graph (Figure 6.7). The following table illustrates this load distribution scheme under the assumption that an adequate number of processors is available: $p = j_{max} + 1$, where j_{max} is the number of defect correction iterates.

processor	calculation of	
1	ζ^0	on the whole interval $[T_0, T_N]$
2	ζ^1	on the whole interval $[T_0, T_N]$
3	ζ^2	on the whole interval $[T_0, T_N]$
$\vdots$	$\vdots$	
$p = j_{\max} + 1$	$\zeta^{j_{\max}}$	on the whole interval $[T_0, T_N]$

If fewer than $j_{\max} + 1$ processors are available, a number of "neighbored" defect correction iterates can be grouped[15] and assigned to one processor. An example of such an arrangement is shown in the following table:

processor	calculation of	
1	ζ^0	on the whole interval $[T_0, T_N]$
	ζ^1	on the whole interval $[T_0, T_N]$
2	ζ^2	on the whole interval $[T_0, T_N]$
	ζ^3	on the whole interval $[T_0, T_N]$
$\vdots$	$\vdots$	
	$\zeta^{j_{\max}-2}$	on the whole interval $[T_0, T_N]$
$p < \lfloor \frac{j_{\max}+1}{2} \rfloor$	$\zeta^{j_{\max}-1}$	on the whole interval $[T_0, T_N]$
	$\zeta^{j_{\max}}$	on the whole interval $[T_0, T_N]$

If the number p of processors is not a divisor of $j_{\max} + 1$, the scheduled order of the IDeC method, different groupings of the defect correction calculations are possible. To find good combinations with a favorable run time performance of the respective parallel IDeC method, it is important to keep load balancing aspects in mind.

The grouping problem is demonstrated further by means of an example with 9 defect correction steps and 4 processors:

---

[15]This grouping corresponds to an assignment of adjacent lines of the dependency graph in Figure 6.7 to *one* processor.

processor	possibility 1	possibility 2
1	ζ^0, ζ^1	ζ^0, ζ^1
2	ζ^2, ζ^3	ζ^2, ζ^3
3	ζ^4, ζ^5	$\zeta^4, \zeta^5, \zeta^6$
4	$\zeta^6, \zeta^7, \zeta^8, \zeta^9$	$\zeta^7, \zeta^8, \zeta^9$

Under the assumption that each ζ^j requires the same amount of calculations and thus needs the same time, possibility 2 is superior in this example. The total run time of possibility 2 is shorter because the load (as represented by the calculation of the iterates $\zeta^0, \ldots, \zeta^9$) is distributed more uniformly.

The assumption that the execution of the basic method (producing ζ^0) and all calculations associated with one defect correction iterate (yielding one of the quantities $\zeta^1, \ldots, \zeta^{j_{max}}$) constitute a similar workload is usually *not* satisfied, especially if ζ^0 is calculated under the influence of a step size control mechanism. A useful load balancing scheme has to take into account factors like this. One possibility is to distribute the workload such that the processor responsible for the basic method is assigned at most as much work as entrusted to the other processors. This request is satisfied if the calculation of the quantities

$$\zeta^0, \ldots, \zeta^{\lfloor j_{max}+1/p \rfloor - 1}$$

is assigned to one processor and all other calculations are distributed as uniformly as possible among the remaining processors.

Figure 6.9 shows the time process diagram (Gantt Chart) of the start-up of an $IDeC_{order}$ method with four defect corrections running on a five processor machine:[16]

- Hatched *rectangles* represent basic steps, and rectangles filled with other patterns represent defect correction steps on one interpolation interval.

- *Arrows pointing upwards* symbolize the transmission of data. First, a processor writes the data into a queue (in Figure 6.9 this queue is assumed to be of length 1). Then it resumes work without delay. When the transmission of data is temporarily impossible (because of a full queue), the processor has to wait until the queue is capable of receiving data again, i.e. until some other processor has fetched the message from the queue. This situation is indicated by diagonal message arrows.

---

[16]The calculation steps of Figure 6.9 are identical in their time requirements with those of the diagram in Figure 6.8, thus making possible a direct comparison.

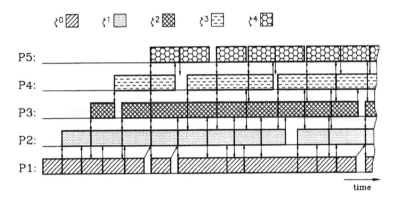

FIGURE 6.9: $\text{IDeC}_{\text{order}}$ – Time Process Diagram

- *Arrows pointing downwards* symbolize the reception of data. They characterize the moment at which a processor collects data from the respective queue.

At the beginning of the calculations on some interpolation interval, every processor (except P1) receives a message from the processor responsible for the preceding interpolation interval. This message contains the approximate solutions of the basic method and the last defect correction values in this interpolation interval. With these initial data the neighboring problem is defined and solved, and the defect correction is carried out. As long as this data is not available, the processor has to remain idle. After a working period, the newly determined data, plus the data from the basic method, is transmitted to the processor responsible for the next defect correction iterate (there is no need to send data for the last processor P5). This procedure automatically ensures the synchronization of the processors involved.

6.6.1   Pros and Cons of $\text{IDeC}_{\text{order}}$ Methods

Advantages:

- Under the assumption that $t_{\mathcal{M}^b} < t_{\mathcal{M}}$, step size rejections do not necessarily lead to idle periods of processors assigned to defect correction tasks. Assume that the processor P1 associated with the basic method is able to produce results faster than the other processors. Its results are put into a queue. After finishing with a certain number of interpolation intervals there is enough time for the processor P1 to recalculate a step after a step size rejection without slowing down the other processors.

- Stochastic run time fluctuations can be levelled out to a certain degree due to the usage of queues (see Figure 6.9).

- IDeC$_{order}$ methods are potentially the fastest if $p = j_{max} + 1$ processors are utilizable.

Disadvantages:

- Larger amounts of data have to be communicated as compared to other parallel IDeC methods. For instance, about $2(M + 1)$ times the amount of the respective data of IDeC$_{time}$ methods has to be transferred.

- If the number of available processors is not a divisor of $j_{max} + 1$ the workload cannot be distributed evenly among these processors. A load imbalance results inevitably in such situations.

- IDeC$_{order}$ methods are more complicated to implement than IDeC$_{time}$ methods.

On account of their properties IDeC$_{order}$ methods can be efficiently implemented only on parallel computers with fast inter-processor communication and the number p of utilizable processors being exactly a divisor of $j_{max} + 1$ (the order of the IDeC method). Shared memory machines meet the communication requirements of IDeC$_{order}$ methods quite naturally. Moreover, the rather complex implementation of IDeC$_{order}$ methods is somewhat easier on this type of parallel computer (Karg et al. [167]).

6.7   IDeC Variant with Adaptive Parallel Defect Corrections

In general, no single method will be most efficient in all situations; rather, different methods will be needed to efficiently handle different cases. Ideally, a parallel IDeC algorithm will choose the most advantageous (most efficient) load distribution itself, making its choice dependent on parameters of the computational environment (computer architecture, number of available processors, communication time between processors, etc.). Such *architecture adaptive algorithms* (AAAs, Krommer, Ueberhuber [7]) decide which algorithm or algorithmic parameters to choose from given sets. Examples of static (a priori) decisions can be found in Augustyn, Ueberhuber [130] and Augustyn, Ueberhuber [133]. The heuristics used for this decision are usually obtained in an empirical way based on practical experiments. However, even the best heuristics will fail if the underlying conditions and assumptions change during the program's execution, e.g., if communication traffic varies.

The basic idea of another approach is that the decision which intervals or which basic/correction steps are calculated next is made dynamically during the program's execution. Algorithms of this type are strongly influenced by the dependency graph described in Augustyn, Ueberhuber [134]. A two-dimensional array is introduced which contains a status number for each interval and for each step (both basic step and correction steps). These status variables are initialized in the following way:

- The status of the first step (i.e. the basic step) of the first interval is given a value of 2.

- The status of the first step of all other intervals are given a value of 1.

- The status of all defect correction steps of the first interval is set to 1.

- All other status variables are set to 0.

Any step of any interval can be started if the corresponding status variable has the value 2, in which case the value of that status variable is immediately increased to 3, to prevent another processor from calculating the same interval again. After the termination of any step, both the status variable of the next correction step (of the same interval) and the status variable of the next interval (of the same correction step) are increased by 1.

During its execution the algorithm must decide which steps are to be calculated next. To accomplish this the following heuristic is used:

- Let k be the number of the highest correction step that can be started (i.e. its status variable is 2). An *early interval* is defined to be an interval that has not yet been calculated and whose correction step number is significantly smaller than k.[17] Such intervals, if found, are calculated first.

- If there are *no* early intervals, then correction steps with low indices should be calculated first. This means that the basic step of each interval receives a higher priority than any of the correction steps.

---

[17]Experiments have shown that an interval should only be processed if k minus its correction step number is 4 (or higher).

Chapter 7

Performance Modelling and Evaluation of Parallel IDeC Methods

7.1 Assessment of Parallel Algorithms

There are many possibilities for devising parallel IDeC methods. A decision
has to be made as to which of the eligible possibilities is most suitable for a
given class of IVPs of ODEs in a specific advanced computing environment.
As a basis for that decision, information is required about the performance of
different IDeC variants in a wide range of computing environments and classes
of ODE IVPs.

Two aspects will be dealt with in this section: (i) what kind of information
is the best foundation for an assessment of the performance of different parallel
IDeC variants, and (ii) how to obtain this information.

7.1.1 Principles of Algorithm Assessment

There are three fundamentally different approaches for evaluating algorithms:

Theoretical Analysis: Algorithms can be characterized by theoretical (i.e.
mathematical) means such as their computational complexity.

   Advantages:

   - Under the inevitable restrictions of such investigations, these meth-
     ods allow precise performance modeling and prediction. It is pos-
     sible, for instance, to obtain a functional description of important
     performance indices. Mean values or upper and lower bounds for
     run time or speed-up can be derived.

Disadvantages:

- Mathematical investigations of parallel algorithms are often very difficult, and sometimes even impossible, to carry out. They may be unfeasible even for moderately complex problems.

- To obtain any results at all, it is often necessary to choose a very high level of abstraction (like in PRAM models) or to make very restrictive assumptions which impair the utilization of the results.

- Comparison of different variants of an algorithm is only possible if they are *all* investigated using the same methodology (on the same level of abstraction, making comparable assumptions, etc.).

Implementation and Testing: All suitable variants are completely implemented and then evaluated on the basis of experimental data.

Advantages:

- No simplifying assumptions are needed. The actual program behavior can be examined on the real target machine(s).

- Direct comparisons of algorithm variants are possible with regard to their run time behavior.

Disadvantages:

- It is not possible to really *understand* the performance of tested programs without models which allow a performance prediction with respect to relevant environmental or problem specific factors.

- Even if just *one* separate performance aspect of an algorithm (such as its communication delays) is to be examined and evaluated, the *whole* algorithm has to be implemented.

- The assessment and comparison of a large number of algorithmic variants requires an enormous, and often prohibitive, programming and testing effort.

- Only single figures of merit (maybe complemented with some kind of statistical characterization) can be acquired. Bounds or limits for performance indices can never be obtained by this type of empirical experimentation alone.

- An exhaustive investigation of algorithms or algorithmic variants is not possible due to the multitude of environmental features of current parallel systems. Evaluations based on implementation and testing can obviously be carried out only on machines available to the test person(s).

- It is impossible to forecast the performance of contemporary algorithms on future parallel systems.

Simulation: First, the performance of the algorithms or algorithmic variants to be evaluated is modeled, and then the resulting models are investigated in planned experiments.

Advantages:

- It is possible to restrict the assessment procedure to features of the examined algorithms which are the center of interest.

  The more realistic a model is (i.e. the lower its level of abstraction is chosen), the better its consistency with the behavior of the actual algorithm will be. In an extreme case, the model may be identical to the fully implemented algorithm. To derive advantage from a model compared to a full implementation of an algorithm, it is important to develop the model on an optimum level of abstraction with respect to the goal of the assessment.

- It is possible to examine aspects that offer unsurmountable resistance to mathematical investigations and that require too much personal and/or computational effort in an investigation based on implementation and testing.

- If all circumstances which influence certain limits or bounds of performance indices (like upper bounds for the speed-up) are known in advance, even such quantities can be derived in simulation studies. Information about these circumstances can be incorporated into the model and may thus permit the evaluation of the respective bound or limit. It is not possible to proceed in this way with the implementation and testing strategy. For instance, the shortest possible communication delays necessary for estimating maximum run time performance cannot be guaranteed in the implemented algorithm.

Disadvantages:

- It is usually difficult and expensive to carefully validate models.[1] To retain the advantage of modeling and simulation as compared with implementing and testing it is not sensible to compare *every* aspect of a model with the behavior of actually implemented algorithms.

- Simulation of massive parallelism usually requires a substantial computational effort on sequential computers.

---

[1]Note the difference between *validation* and *verification*. Verification is an activity to ensure the correspondence between the model and its specification. Validation on the other hand aims to establish the consistency between a model and reality. Validation is usually much more difficult than verification.

- To obtain models which agree with the real algorithm's behavior and which are as simple as possible ("Occam's razor"), it is necessary to make simplifying assumptions. These abstractions and simplifications are normally not as restrictive as those needed in purely theoretical investigations.

7.1.2   Performance Evaluation Criteria

Performance assessment is the goal of the examination of parallel IDeC algorithms as discussed in the following. The question to be investigated is the following: how good is the absolute and relative performance of certain instances of the family of IDeC algorithms (possibly in comparison with other algorithms) under various environmental and problem specific conditions? Such an evaluation or comparison can be based on various criteria: run time behavior, hardware requirements, personal effort needed for the algorithm and software development, overall costs, etc.

As the following discussion is centered on scientific aspects, the run time behavior will be the primary performance criterion. However other features of the investigated algorithms will not be left out of consideration.

The IDeC variants studied in the following sections differ primarily in their allocation of parallel tasks to physical processors. They are more or less indistinguishable with respect to their numerical properties (stability, round-off error effects, etc.). Therefore numerical aspects will not be included in the following evaluation of parallel IDeC methods.

Performance Indices

Various indices can be used to characterize the performance of scientific programs on advanced – in particular on parallel and distributed – computer systems:

Absolute Performance Metrics are based on measured, simulated, or theoretically predicted run times and on the knowledge of the required number of floating-point operations. For instance (Addison et al. [127]):

> **Temporal Performance** $R_T(N;p) := T^{-1}(N;p)$ is the reciprocal run time of the parallel algorithm; N denotes the problem size and p the number of processors used. The unit of this metric is *time steps/s*.
>
> This performance index reaches a maximum when the algorithm's run time is at a minimum, no matter how many floating point operations are actually carried out.
>
> **Benchmark Performance** $R_B(N;p) := F_B(N)/T(N;p)$ is the ratio between the number of floating-point operations $F_B(N)$ needed for

the execution of an equivalent sequential algorithm and the run time $T(N;p)$ of the equivalent parallel algorithm. The unit of this metric is *flop/s* (or *Mflop/s, Gflop/s*).

Hardware Performance $R_H(N;p) := F_H(N)/T(N;p)$ is the ratio between the required number of floating point operations $F_H(N)$ of the parallel algorithm and its run time. The unit of this metric is also *flop/s* (or *Mflop/s, Gflop/s*).

Relative Performance Metrics are derived by comparing the performance of a parallel algorithm to certain kinds of appropriately chosen performance standards, such as the performance of an equivalent sequential algorithm or program.

Speed-up $S_p := T_1/T_p$

Speed-up is the ratio of the run time T_1 of the sequential algorithm (prior to parallelization) to the run time T_p of the parallelized algorithm utilizing p processors.

Efficiency $E_p := S_p/p \leq 1$

Concurrent efficiency is a metric for the utilization of a multiprocessor's capacity. The closer E_p gets to 1, the better use is made of the potentially p-fold power of a p processor system.

Effectivity $F_p := S_p/(pT_p) = E_p/T_p = E_pS_p/T_1$

Effectivity is the ratio between speed-up and work as characterized by pT_p. It measures both the absolute increase of speed *and* the relative utilization of the computer's capacity. An effective algorithm maximizes F_p. With the aid of F_p or F_pT_1 it is possible, for instance, to determine the most favorable number of processors to be employed.

System Efficiency is an interesting relative performance index obtained from relating benchmark performance data and/or hardware performance indices with the theoretical peak performance of a machine, i.e. its maximum rate of execution based on the cycle time of the hardware.

Absolute performance indices allow direct comparisons of different algorithms or computer architectures (Addison et al. [127]). However, their application is sometimes problematic, as these metrics are constructed primarily for benchmarks, i.e. for the performance assessment of existing programs on a given hardware assessment. With the exception of R_T, it is necessary to know the number of floating point operations performed *and* the required run time in order to calculate these indices. In particular, the number of executed floating point operations executed is difficult to obtain in practical numerical data processing. For instance, the integrand functions in numerical integration programs are usually "black boxes". No a priori statement can be made regarding the number of operations executed in them.

Speed-up S_p and efficiency E_p are standardized (relative) performance indices based on run time ratios; thus they avoid the inherent difficulties of absolute metrics. In contrast to absolute metrics, these two metrics, together with the effectiveness F_p which is derived from them, also offer information on the operating costs of algorithms.[2] The system efficiency of a program is an indicator of the degree of utilization of the underlying computer system.

In all cases where an unbiased comparison is not possible because of different sequential run times, the absolute temporal performance R_T can be used as an alternative metric.

In the following chapters, speed-up S_p, efficiency E_p, effectiveness F_p, and the temporal performance R_T will be used for an assessment of the investigated IDeC variants. All these indices can easily be determined either by experiments or by simulation.

The reader should be careful when comparing the results of the following chapters with performance data of other algorithms (published elsewhere).

7.1.3   Simulation of IDeC Methods

Investigating the influence of environmental factors and specific problem characteristics on the performance of parallel IDeC algorithms in a purely mathematical analytical type of examination is out of the question. The degree of complexity of the system to be examined is too high (in particular because of the queues involved).

On the other hand, if parallel IDeC methods are to be implemented and compared, many numerical aspects have to be taken into consideration. This is especially costly for implicit variants.

Simulating parallel IDeC variants is obviously the best way to examine their performance characteristics. To simulate the run time behavior of parallel IDeC methods with C-programs on a sequential computer is a straightforward matter (Augustyn, Ueberhuber [131]). The resulting simulation programs are not restricted to IDeC methods. They can be used to examine other parallel algorithms as well, provided suitable models for these algorithms are constructed.

The model of the IDeC methods which was investigated with the simulation software mentioned above is based on the following specific assumptions:

- Solving the complete IVP requires a *fixed* number K of interpolation intervals $[T_i, T_{i+1}]$, $i = 0, ..., K-1$. The quantity K can be set as a pa-

---

[2]Two algorithms with identical run time behavior can still differ in their operating costs, for example, on account of different numbers of processors required to achieve a specific performance. Information of this type can be obtained from S_p and E_p, but not from R_T.

rameter of the simulation, i.e. all examined IDeC variants are assumed to perform the same sequence of step sizes.

Only with a fixed K is it possible to carry out performance examinations and comparisons which are independent of the properties of a specific initial value problem.[3] This simplifying assumption does not severely restrict the generality of these examinations.

- Run times $t_{\mathcal{M}^b}$ for the basic method $\mathcal{M}^b$ and $t_{\mathcal{M}}$ of the method $\mathcal{M}$ applied to one neighboring problem are simulated under the following circumstances:

  - It is assumed that the entire IDeC method is controlled by some *local error estimation* mechanism linked to the basic method $\mathcal{M}^b$.

    The investigated parallel IDeC methods are all based on the global connection strategy (Augustyn et al. [129]). Therefore numerical results provided by the defect correction iterates can only be used to estimate *global errors*. However, controlling the solution of IVP ODEs by global error estimation is unfeasible (Shampine, Watts [181]).

    Controlling the IDeC process over the basic method through a step size mechanism independent of the defect correction iterates seems to be the best solution.

  - The *average* run time $t_{\mathcal{M}^b}$ for the solution of the basic method $\mathcal{M}^b$ is assumed to be fixed. It is affected, however, by uniformly distributed stochastic disturbances. With a certain probability (assumed to be time invariant) a step of the basic method is rejected. The rejection of a step may be repeated up to three times[4], and thus the calculation time may be increased by up to a factor of three.

  - The average calculation time $t_{\mathcal{M}}$ for one defect correction is assumed to be identical for all correction steps.[5] Every single correction step is affected by a uniformly distributed stochastic disturbance in the same way as the basic step.

---

[3]In practice K is normally not chosen a priori but determined by a step size control mechanism which adapts the interpolation intervals to the given IVP.

[4]It seems sensible to limit the repetition of steps to three times. Should a step be rejected even more often (for instance, because of a singularity) it is obvious that some special treatment is required for this "trouble spot". Special cases of this type are not treated in this book.

[5]In the following $t_{\mathcal{M}^b}$ or $t_{\mathcal{M}}$ will describe the complete calculation time for a basic or correction step and not only for the underlying procedure for solving a system of differential equations.

- Every processor requires a certain amount of time for communication purposes before and after each step. For the sake of simplicity this time period is assumed to proceed without stochastic fluctuations. It is also assumed that the communication delay for a given message is the same regardless of *which* two processors are communicating with each other.

  Data transmission between two processors can be characterized approximately by a linear model with two parameters:

  1. The *latency* of the data transmission is the theoretical run time of a message of length zero and is thus independent of the length of the message. Essentially, the latency period is determined by the communication protocol used and the signal run time on the data-line. This time period can vary greatly for different parallel computers.

  2. The specific *data transmission time* is required to send a message of unit length via the transmission media (without start-up time).

  This simple model permits only rough estimates of communication times depending on the length of the messages[6] , but is sufficiently accurate for an assessment and comparison of parallel IDeC-variants.

  A further assumption of the simulation is the separation without overlap of the transmission and reception periods. Only this form allows the use of queues for the transmission of messages. A transmitter does not have to wait for a processor ready to receive messages. Nevertheless a total amount of data transmission time can be given (sum of transmission and reception time periods).

- Communication between processors is established via FIFO queues of a certain maximum length.

- Synchronization and administration overhead is neglected.

The goal of this simulation study was to determine IDeC variants which achieve satisfactory speed-up and/or efficiency under certain environmental conditions and specific problem characteristics. In this sense the following parameters have been chosen as a basis for the respective investigations:

---

[6]In reality transmission times are affected by stochastic uncertainties. Besides, discontinuities of transmission time can occur depending on the length of message.

- number p of processors;
- number $j_{\max}$ of defect correction iterates;
- number K of interpolation intervals;
- $t_{\mathcal{M}b}/t_{\mathcal{M}}$ ratio;
- run time fluctuations[7];
- communication delays[8];
- length of the queues assumed for the data exchange between processors.

7.2 IDeC Variant with Parallel Time Steps

The run time performance of various IDeC$_{\text{time}}$ methods was investigated by simulation on a sequential computer (Augustyn, Ueberhuber [131]). The underlying performance models and the obtained simulation results were validated on different parallel computers (Karg et al. [168, 170]).

Figures 7.1 and 7.2 show simulation results of an tenth order IDeC$_{\text{time}}$ method (nine defect correction iterates).

---

[7]These fluctuations may originate, for instance, from the specific feature of implicit methods to require a varying number of iterations for the solution of their associated systems of algebraic equations. Additionally the run time of the basic method depends strongly on the step size control mechanism. The probability of rejecting already calculated steps has been taken into account in the simulation study.

[8]The time needed to communicate numerical data between the processors of a parallel machine.

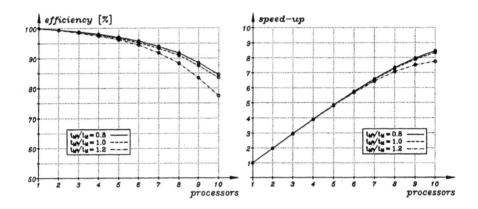

FIGURE 7.1: IDeC$_{time}$ – Influence of the Number of Processors

Simulation parameters for Figures 7.1 and 7.2:

degree M of the polynomials	10	average ratio t_{M^b}/t_M	0.8, 1, 1.2
j_{max} defect correction iterates	9	fluctuations of t_{M^b}	30 %
number p of processors	*variable*	fluctuations of t_M	30 %
K interpolation intervals	1000	communication delay	0
probability of step rejection	10 %	length of queue	–
simulation runs	10		

The decline of the efficiency when the number of processors goes up can be explained as follows:

- The start-up process of IDeC$_{time}$ methods utilizing a larger number of processors requires more idle periods (see Figure 6.8). This leads to an efficiency degradation when the number of processors is augmented. The same phenomenon occurs with the shutdown process at the end of the integration interval.

- The situation $t_{M^b} > t_M$ leads to idle times at all defect correction levels because the basic step has to be finished before any defect correction calculation can be started on the corresponding interpolation interval. This results in a performance degradation. The small performance loss still present in situations with $t_{M^b} \leq t_M$ is caused by the step size mechanism which also introduces idle periods. This effect can be understood when

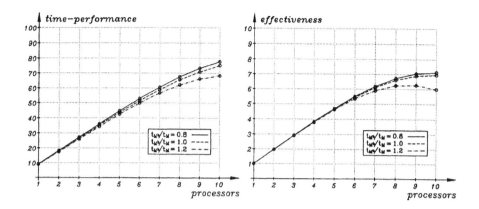

FIGURE 7.2: IDeC$_{\text{time}}$ – Influence of the Number of Processors

looking at the Figure 6.8 – just suppose some of the rectangles associated with the basic method (calculating ζ^0) are enlarged.

- As long as the number p of processors is not larger than $j_{\max} + 1$ there is enough time for the processors concerned with defect correction iterates to finish their calculations before the respective final value is required by another processor as a starting value. The probability that a processor has to wait for its required starting values increases with the utilized number of processors. This also contributes to a deteriorated performance when a larger number of processors is used.

The product of speed-up and efficiency gives a performance index which is proportional to the effectiveness F_p defined in Section 7.1.2. $F_p T_1 = S_p E_p$ will be called *effectiveness* in the following, simplifying the calculations without loss of generality. The effectiveness of IDeC$_{\text{time}}$ methods is shown in Figure 7.2 together with the temporal performance R_T inversely proportional to the run time.

In the case of $t_{\mathcal{M}^b} > t_{\mathcal{M}}$ the effectiveness reaches a maximum when the number of processors is still smaller than $j_{\max} + 1$. However, the shortest overall run time is achieved when the maximum number of useful processors is used, at the expense of not fully utilizing the available capacity of the processors. The time gained in comparison with an IDeC method operating at maximum effectiveness is rather small.

Figure 7.3 shows the dependence of efficiency, speed-up and effectiveness on the order of the IDeC method (the number of defect correction iterates plus 1) and the number K of calculated integration intervals. These curves are derived using the maximum useful number of processors $(p = j_{max} + 1)$ in order to achieve a run time minimum.

Simulation parameters for Figure 7.3:

degree M of the polynomials	*variable*	average ratio t_{M^b}/t_M	0.8
j_{max} defect correction iterates	$M - 1$	fluctuations of t_{M^b}	30 %
number p of processors	M	fluctuations of t_M	30 %
K interpolation intervals	10, 100, 1000	communication delay	0
probability of step rejection	10 %	length of queue	–
simulation runs	10		

The more integration intervals calculated, the more pronounced the independence of the utilization of the processor's capacity from the order of the IDeC method. The drop in efficiency at the transition from one to two (and more) processors is a result of statistical run time fluctuations, the start-up process and the shutdown period near the end of the integration interval. These effects do not take place in a single processor setup.

The decline of efficiency when the number of integration intervals decreases is the result of a stronger influence of the start-up and shutdown periods. The higher the order chosen, the longer these periods are. The start-up of an IDeC$_{time}$ method is graphically portrayed in the Gantt Chart of Figure 6.8.

Figure 7.3 shows an increase of effectiveness with increasing order and number of integration intervals. A maximum shows only when K is very small, i.e. start-up and shutdown periods play a dominant role. This maximum is very flat and lies near an order of 10. In practice, orders higher than 10 are normally not very useful. A conclusion drawn from the simulation results is that high order IDeC methods are preferable. The higher the order chosen, the more parallel processors can be utilized.

Figure 7.4 shows the effect of stochastic run time fluctuations for different time ratios t_{M^b}/t_M. A pronounced decrease in the utilization of the available processors' capacities and in the respective run time performances R_T can be observed as soon as the run time t_{M^b} of the basic method is greater than the run time t_M of one defect correction iterate. It is obvious from these diagrams that run time optimization and further parallelization measures should primarily concentrate on the basic method, as this should lead to the greatest increase of speed. Further, run time fluctuations have a striking effect on the total run time of IDeC$_{time}$ methods for $t_{M^b} < t_M$.

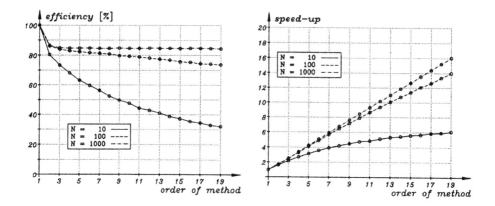

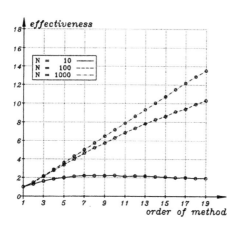

FIGURE 7.3: IDeC$_{time}$ – Influence of the Order and the Number of Subintervals

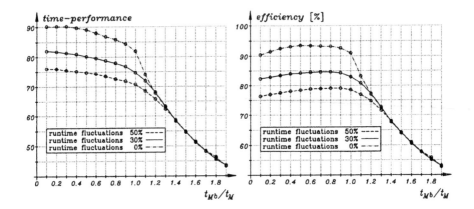

FIGURE 7.4: IDeC$_{time}$ – Influence of Run Time Fluctuations

Simulation parameters for Figure 7.4:

degree M of the polynomials	10	average ratio t_{Mb}/t_M	*variable*
j_{max} defect correction iterates	9	fluctuations of t_{Mb}	0 %, 30 %, 50 %
number p of processors	10	fluctuations of t_M	0 %, 30 %, 50 %
K interpolation intervals	1000	communication delay	0
probability of step rejection	10 %	length of queue	–
simulation runs	5		

Since in this case efficiency, speed-up and effectiveness differ only by a constant factor only one of these indices (efficiency) is given. Interpreting this diagram, the total run time seems to *increase* for decreasing t_M. This contradiction can be resolved when the different sequential run times used to calculate the portrayed performance index are taken into account (sequential run times are different for each t_{Mb}/t_M ratio). With the run time performance R_T no such problem is encountered, showing an important advantage of R_T as a performance index. This example shows how much care is needed when interpreting *relative performance indices*.

7.2.1 Validation

In order to validate the simulation results of IDeC$_{time}$ methods the parallel parts of the algorithm were implemented on several parallel computers. In

these studies some sequential parts of the defect correction process and the
step size control mechanism were replaced by suitable models analogous to
those of the simulation program. A full implementation of the method would
have been too expensive and too difficult to evaluate to be useful (as discussed
in Section 7.1.3).

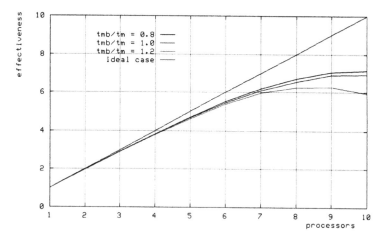

FIGURE 7.5: IDeC$_{time}$ Method on a Shared Memory Architecture

Both main types of parallel computers, shared and private memory systems,
were used in the validation studies: primarily (i) a Sequent Balance with 26
processors (shared memory architecture) and (ii) an iPSC/860 Hypercube with
16 processors (distributed memory architecture).

Detailed results of the validation studies for IDeC$_{time}$ methods may be found
in two reports (Karg et al. [168, 170]). The simulation results are remarkably
accurate compared with the behavior of the implemented parallel programs.
For example, Figure 7.5 shows the effectivity of an IDeC$_{time}$ program on the
Sequent Balance. A comparison with Figure 7.2 reveals an astonishing agreement of the simulated and the actual behavior. When interpreting the figure, it
should be taken into consideration that the simulation of IDeC$_{time}$ methods as
presented in this book does not cover the influence of different communication
delays. This subject is treated in detail in (Augustyn, Ueberhuber [132]).

7.3   IDeC Variant with Parallel Defect Corrections

The run time behavior of IDeC$_{order}$ methods was investigated in a simulation study on a sequential computer as described in Chapter 7.1.3 (see also Augustyn, Ueberhuber [131]). Similarly, the models and the corresponding simulation results were validated on various parallel computers (Karg et al. [167, 169]).

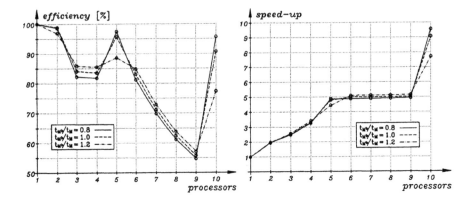

FIGURE 7.6: IDeC$_{order}$ – Influence of the Number of Processors

Figure 7.6 shows the efficiency and the speed-up of IDeC$_{order}$ methods in relation to the number of utilized processors p. The seemingly strange curves are the result of grouping several defect correction iterates for $p < j_{max} + 1$. Local efficiency maxima occur when p evenly divides into $j_{max} + 1$. If p is not a divisor of $j_{max} + 1$, the workload is unevenly distributed among the processors. In these cases the utilization of the potential power of the available processors is far from optimal.

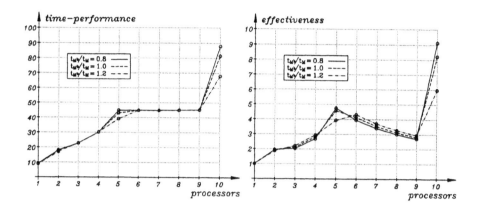

FIGURE 7.7: IDeC$_{order}$ – Influence of the Number of Processors

Simulation parameters for Figures 7.6 and 7.7:

degree M of the polynomials	10	average ratio $t_{\mathcal{M}^b}/t_{\mathcal{M}}$	0.8, 1, 1.2
j_{max} defect corrections iterates	9	fluctuations of $t_{\mathcal{M}^b}$	30 %
number p of processors	*variable*	fluctuations of $t_{\mathcal{M}}$	30 %
K interpolation intervals	1000	communication delay	0
probability of step rejection	10 %	length of queue	5
simulation runs	10		

For $t_{\mathcal{M}^b} > t_{\mathcal{M}}$ and p not being a divisor of $j_{max} + 1$ a remarkable phenomenon takes place: efficiency and speed-up are better than for $t_{\mathcal{M}^b} \leq t_{\mathcal{M}}$. This effect is the result of assigning the basic method to the processor with the smallest workload. Therefore $t_{\mathcal{M}^b} > t_{\mathcal{M}}$ leads to a better utilization of this processor as long as $t_{\mathcal{M}^b} \leq 2t_{\mathcal{M}}$.[9] Otherwise the run time of processor P1 responsible for the basic method exceeds that of the other processors, causing a rapid degradation of efficiency and speed-up. However, this phenomenon has no practical relevance because IDeC$_{time}$ has a far better performance under these conditions (see Figure 7.2) and should be preferred anyhow.

Run time performance and effectiveness results of these simulation runs are shown in Figure 7.7.

---

[9]The processors' work-load differs at most by *one* defect correction iterate. In the case of p not being a divisor of $j_{max} + 1$, processor P1 has enough spare time to cope with $t_{\mathcal{M}} < t_{\mathcal{M}^b} < 2t_{\mathcal{M}}$ (P1, which is responsible for the basic method, is assigned less work than the other processors in this situation).

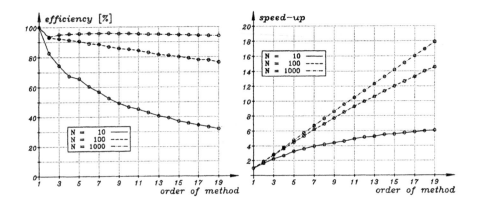

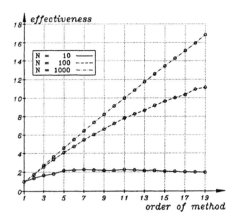

FIGURE 7.8: IDeC$_{order}$ – Influence of the Order and the Number of Subintervals

Obviously $IDeC_{order}$ methods are advantageous only if the number of processors is a factor of the order of the IDeC method. They show a particularly good performance if $p = j_{max} + 1$. All simulation runs for determining the performance indices of Figure 7.8 were based on the assumption $p = j_{max} + 1$, thus giving results only for the best possible $IDeC_{order}$ methods (with regard to the number of utilized processors). The qualitative behavior of speed-up, efficiency and effectiveness is very similar to the respective indices of $IDeC_{time}$ methods. The conclusions drawn in Chapter 7.2 remain valid for $IDeC_{order}$ methods.

Simulation parameters for Figure 7.8:

degree M of the polynomials	*variable*	average ratio t_{M^b}/t_M	0.8
j_{max} defect correction iterates	$M - 1$	fluctuations of t_{M^b}	30 %
number p of processors	M	fluctuations of t_M	30 %
K interpolation intervals	10, 100, 1000	communication delay	0
probability of step rejection	10 %	length of queue	5
simulation runs	10		

The necessary length of the queues and the influence of run time fluctuations can be deduced from the results of further simulation runs as presented in Figure 7.9.

Simulation parameters for Figure 7.9:

degree M of the polynomials	10	average ratio t_{M^b}/t_M	*variable*
j_{max} defect correction iterates	9	fluctuations of t_{M^b}	0 %, **30 %**, 50 %
number p of processors	10	fluctuations of t_M	0 %, **30 %**, 50 %
K interpolation intervals	1000	communication delay	0
probability of step rejection	10 %	simulation runs	10
length of queue	1, **5**, 100		

A queue length of 5 seems to be sufficient for all relevant numbers of processors. More detailed examinations establish the fact that a perfectly satisfactory performance can be maintained if the values of the basic method are transmitted via a queue of length 5 and the data of the defect correction iterates via queues of length 2 or 3.

In contrast to the performance of $IDeC_{time}$ methods, the overall run time of $IDeC_{order}$ methods is almost independent of temporal fluctuations of the defect correction calculations. This is a result of the queues used to level out such variations. As was expected, there is no difference between run time behaviors of $IDeC_{order}$ and $IDeC_{time}$ methods as soon as the run time t_{M^b} of the basic method

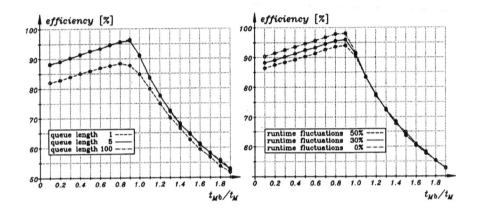

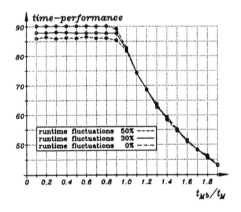

FIGURE 7.9: IDeC$_{order}$ – Influence of Queue Length and Run Time Fluctuations

exceeds the time $t_\mathcal{M}$ needed for the calculations associated with one defect correction iterate.[10] If $t_{\mathcal{M}^b}$ exceeds $t_\mathcal{M}$ the total run time of the IDeC method is determined only by the time needed for the basic method. Thus for $t_{\mathcal{M}^b} > t_\mathcal{M}$ *all* IDeC variants should have more or less the same run times, as long as communication delays do not play a dominant role (see Augustyn, Ueberhuber [132]).

Obviously the most important factor influencing the performance of parallel IDeC methods is $t_{\mathcal{M}^b}$, the run time of the basic method. This holds especially for $t_{\mathcal{M}^b} > t_\mathcal{M}$. Any measure taken to improve the speed of the basic method will immediately decrease the total computation time, as can be seen from the R_T performance in Figure 7.9.

7.3.1 Validation

The validation of the simulation results concerning IDeC_order methods was based on the same methodology as the respective validation of the IDeC_time simulation (as described in Section 7.2.1).

Shared memory systems (a Sequent Balance with 26 processors) and distributed memory computers (an iPSC/860 Hypercube with 16 processors) were used in validation studies.

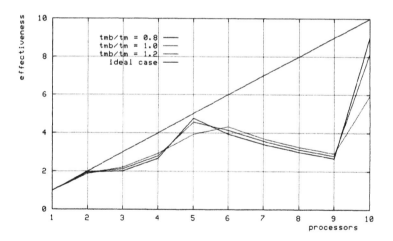

FIGURE 7.10: IDeC_order Method on a Shared Memory Architecture

Detailed results of the validation studies for IDeC_order methods may be found in the publications Karg et al. [167, 169]. Again, the simulation results are remarkably accurate compared to the behavior of the implemented parallel

---

[10]Step size rejection is the cause for decreasing run time even when $t_{\mathcal{M}^b} < t_\mathcal{M}$.

programs. For example, Figure 7.10 shows the effectivity of an $IDeC_{order}$ program on the Sequent Balance. A comparison with Figure 7.7 shows a strong agreement between the simulated and the actual behavior. As mentioned in Section 7.2.1, the simulation does not cover the influence of different communication delays. The influence of this hardware granularity aspect is described in Augustyn, Ueberhuber [132].

Chapter 8

Representative Target Machines

The architecture of the parallel machine has a strong influence on the program structure, which in turn influences the internal proceedings of the programs observed in the visualization. Therefore two machines with contrasting architectures were chosen: (i) a *Sequent Balance*, a shared memory computer, and (ii) an *Intel Hypercube*, a message passing parallel system. Both parallel machines are installed in the Computer Center of the Technical University of Vienna.

8.1   Sequent Balance B 21

The Sequent Balance B 21 has 28 processor nodes which communicate with each other via a 24 MByte shared memory. The operating system DYNIX (a slightly modified UNIX version) is responsible for managing all processors. Processes can be started on arbitrary nodes. Parallelism is achieved by starting multiple processes. The programming language used is C, enriched by the following constructs to express parallelism:

- The identifier `shared` declares that a static variable belongs to the shared memory. The identifier `private` declares that a static variable does *not* belong to the shared memory.

- To allocate shared memory dynamically there exists a library function `sh_malloc` which is the counterpart to the standard `malloc` function.

- Functions are available for setting the number of used processors, and for starting a process on one processor.

- A library that supports semaphores and barriers may be used.

The IDeC programs for the Sequent Balance make extensive use of the shared memory to accomplish interprocess communication. Messages of length one are only used to signal the availability of the data in the shared memory.

8.2   Intel Hypercube iPSC/860

The iPSC/860 has a hypercube architecture. This means that each node k is directly connected with node $z = (k \text{ } eor \text{ } 2^i) \text{ MOD } 2^d, i = 0, \ldots, d - 1$.[1] The used iPSC/860 has 16 processors $(d = 4)$. The processor of each node is an INTEL i860, a RISC processor with integrated floating point unit. Each processor node has 8 MByte local memory. Processors communicate with each other via message-passing – a shared memory is *not* available.

The front-end of the hypercube is a PC running UNIX, which is responsible for managing the available nodes and starting processes on particular nodes of the hypercube. Consequently, software development for the iPSC/860 requires two kinds of programs:

1. First, a program, that runs on the master (the PC) is needed. This program is responsible for the user interface and for calling the necessary functions for allocating nodes on the hypercube and for starting programs on those nodes.

2. The second type of program runs on one or more of the hypercube nodes.

Again, C is used as the programming language. On the master there are functions available that allow the allocation of processor nodes and the initiation of node programs. On each node, functions are available that permit the sending and receiving of messages.

The IDeC programs for the hypercube use messages for interprocess communication. These messages contain all data necessary for the computation (i.e. arrays or matrices, respectively). Consequently they are longer than those of the IDeC programs for the Sequent Balance.

---

[1] *eor* is the bitwise *exclusive-or*: 0 *eor* 0 = 0, 1 *eor* 0 = 1, 0 *eor* 1 = 1, 1 *eor* 1 = 0.

Chapter 9

Trace File

A *trace file* is a list of events, recorded during an execution of the parallel program. This file is usually "human-readable" (i.e. an ASCII file). Each line contains a separate trace record which corresponds to one program event. Events may be the sending and receiving of data, the starting and stopping of program *tasks*[1], etc. Since PARAGRAPH (Chapter 5) demands trace files in PICL format (Portable Instrumented Communication Library; Geist et al. [4]), all investigated parallel IDeC programs have been modified accordingly to produce such trace files.

9.1 PICL

PICL is a *portable instrumented communication library* designed to provide portability, ease of programming, and execution tracing in parallel programs.

PICL provides portability between many machines and multiprocessor environments. It is fully implemented, for instance, on the Intel iPSC/860, the Ncube/3200 families of hypercube multiprocessors and on the Cogent multiprocessor workstation. The authors have announced that this list will grow as soon as new machines and programming environments appropriate for the library appear.

In addition to supplying low-level communication primitives, such as **send** and **receive**, PICL simplifies parallel programming by providing a set of high-level communication routines such as global broadcast, global maximum, and barrier synchronization. These routines can help the novice user avoid common synchronization and programming errors and save programming time even for the veteran user. These high-level routines also facilitate experimentation and performance optimization by supporting a variety of interconnection topologies.

---

[1] *Tasks* in the context of IDeC algorithms are defined in the following way: The basic step is labelled **task 0** and the kth defect correction step is labelled **task k**.

Execution tracing has been built into the PICL routines, and routines are provided to control the type and amount of tracing. The tracing facility is useful for performance modeling, performance tuning, and debugging.

The library is made up of three distinct sets of routines: a set of low-level generic communication and system primitives, a set of high-level global communication routines, and a set of routines for invoking and controlling the execution tracing facility. In its current version, the library is implemented in C; a Fortran version is under development. In the meantime, Fortran-to-C interface routines are provided so that Fortran programs can call the C version of the library.

9.1.1 Low-Level Primitives

On each of the target machines, the user has at his disposal a set of machine-dependent system and communication routines that enable him to use the underlying parallel architecture. Because each of the target machines and programming environments supports the same general message-passing programming paradigm, most of these system and communication functions are available on all of the machines, though the subroutine names and parameter lists vary from machine to machine. Most importantly, this set of common functions and capabilities is sufficient to produce efficient, well-written codes. By providing the user with a generic interface to this common pool of functions, these library routines enable the user to write codes that can be transported without change to any machine on which the library has been implemented. The library contains a generic interface routine for each of the functions described in Table 9.1.

A function or capability available on one target machine but not on another obviously cannot be included in a library intended to to achieve portability for application codes. Thus, PICL supports a restricted programming model that includes only capabilities and functions available on all target multiprocessing environments.

Programming Model

The programming model can be viewed as a generic form of the message-passing programming environments found on most distributed-memory processors. It assumes a set of autonomous processors, each possessing a fixed size memory to which no other processor has access. Processors share data by communicating messages to each other. More specifically, if processor i has data required by processor j, then i must send the data to j by issuing a **send0** command, and processor j must issue a **recv0** command in order to receive the message. From the user's viewpoint, processor i is idle (or *blocked*) from the time it issues the **send0** command until the message is copied from the user's message buffer

open0 :	Open interprocessor communication.
close0 :	Close interprocessor communication.
load0 :	Load a node program.
sync0 :	Synchronize the processors.
send0 :	Send a message.
recv0 :	Receive a message.
probe0 :	Check for an arriving message.
recvinfo0 :	Return information about the most recently received or checked ("probed") message.
message0 :	Print a message on the standard output device.
who0 :	Return processor ID number, host ID number, and number of node processors allocated.
clock0 :	Read the system clock.
check0 :	Enable or disable parameter checking.

TABLE 9.1: Low-Level PICL Routines

into a system buffer, at which time the user's message buffer can be reused safely. Processor j is blocked from the time it issues the recv0 command until a message satisfying the request arrives and is copied into the specified user buffer. Note that the program will never terminate if a recv0 command is never satisfied by an arriving message. The programmer must design his programs so that there is an incoming message corresponding to each call to recv0.

The model also assumes that interprocessor communication is *interrupt-driven*: if processor i calls send0 to send a message to processor j before processor j calls recv0 to receive it, then the incoming message causes processor j's operating system to interrupt whatever task it is currently processing in order to receive the message and store it in a system buffer. When the user's program on processor j finally issues a request for the message, the message is copied from the system buffer to the buffer provided by the user's program. Thus, PICL supports an *asynchronous programming* style, rather than a synchronous style where each sending processor blocks until the receiving processor has issued the corresponding recv0.

PICL makes no assumptions about the underlying communication network and relies on the ability of the target multiprocessors to send messages between arbitrarily chosen pairs of processors. The time required to send a message between two processors will often be a function of the interprocessor communication network, and a user will need to be aware of such machine dependencies in order to write efficient programs.

The model distinguishes one processor, the *host*, from the rest. The user has access to the remaining processors, the *node* processors (or simply nodes), *only* through the host. Typically, an application code consists of one program that runs on the host and another program (or set of programs) that run(s) on the node processors. The host program calls system primitives to allocate node processors, load the node program (or programs) on the nodes, send input data required by the node programs, and receive results from the nodes.

Restrictions

While the library implements a somewhat limited programming model, it more than adequately meets most users' needs, as there cannot be found excluded features and capabilities essential in parallel programs. The following features, available on some target multiprocessors, are *not* supported:

1. PICL does not support running more than one process (or node program) simultaneously on a single processor.

2. The use of *nonblocking* interprocessor communication is not supported.

3. PICL does not support user-defined handlers for system interrupts.

4. PICL does not explicitly support *synchronous* communication.

Although users whose application programs require capabilities not supported by the library will not be able to use this library to achieve full portability, they can still use it to reduce the amount of code they have to write and the number of changes required when porting a program to another machine.

Portability

The use of generic interface routines that in turn call vendor-supplied native routines raises the following fundamental portability issue. Each of the target multiprocessors has its own range and interpretation of legal parameter values. But, confining the user to universally valid parameter values is too restrictive, and it would become more restrictive as the number of programming environments on which the library is implemented grows. It was therefore decided to allow the user to employ the full range of valid parameter values on each target machine, with the exception of a small subset of values reserved for use by the library itself. On each machine, the routines perform conventional error trapping, checking each input parameter to see if it falls in the range of valid values for that particular machine. When an invalid value is encountered by a routine, an error message is sent to the standard output device of the host and the routine stops processing. Thus, while the user is required to select parameter values valid on each of his target machines, when he mistakenly chooses a value that causes an error condition on one of the machines, his program fails gracefully, letting him know precisely what happened.

`traceenable:`	Enable tracing and specify the name of the trace file.
`tracehost:`	Begin tracing on the host.
`tracenode:`	Begin tracing on a node processor.
`tracelevel:`	Specify the amount and type of trace information.
`traceinfo:`	Return the current `tracelevel` specification.
`tracemark:`	Generate a user-typed trace record.
`tracemsg:`	Write a line of text directly into the trace file.
`traceflush:`	Send trace information to the trace file *now*.
`traceexit:`	Stop tracing.

TABLE 9.2: PICL Routines for Tracing

9.1.2 High-Level Communication Routines

The routines included in the high-level communication library are those that have proven most useful in the development of parallel algorithms and application programs for distributed-memory machines. The underlying topology to be used by the high-level communication routines may be specified. In this way broadcasting messages may be optimized, processors may be synchronized, and various vector operations (i.e. functions where several processors are involved) may be computed efficiently.

However, these high-level communication routines are *not* used by the IDeC programs, and will therefore not be discussed further.

9.1.3 Execution Tracing

When the user requests execution tracing, he activates code within the low-level primitive and/or high-level communication routines that produces time-stamped records detailing the course of the computation on each processor. The information it produces can be used, for instance, by PARAGRAPH to help in analyzing performance or debugging the code. One of the key quantities captured is a record of the time each processor spends blocked while waiting for messages from other processors. With this and similar data the user can evaluate the performance of his code and locate possible performance bottlenecks. Execution tracing is controlled by the routines described in Table 9.2.

It is crucial that the tracing facility have minimal impact on the performance of the code being studied. Three features of PICL contribute to that goal.

- The tracing is "event-driven": trace information is generated only when an event of interest has occurred. The default events are sends, receives, and other events connected with interprocessor communication. Thus, tracing is guaranteed to have minimal impact on stretches of computation in which little interprocessor communication occurs.

- The user can control the amount and type of trace data that is collected. By using the `tracelevel` command to enable and disable the various types of tracing and to control the level of detail recorded, he can collect detailed data only where needed and avoid generating extraneous data from portions of the code not currently of interest.

- The tracing facility stores trace records into a user-specified block of internal memory. There is no external communication or I/O associated with tracing as long as this internal trace array is not filled to capacity before the computation is complete. Node trace records are then automatically sent back to the host after *all* node programs have completed all other processing.

While the approach of PICL is inevitably more intrusive than a performance monitor built into the operating system (perhaps with hardware support), such a monitor is not available on most machines and, in any case, would be highly non-portable. Experience with the trace facility indicates that it generally has a small effect on the performance of the code being instrumented, but not enough to change the important features of its run time behavior.

9.2 Trace File Generation

The generation of trace files for the Intel Hypercube is simple. Since this parallel architecture is directly supported by PICL, the program code needs only a few additional commands to produce an adequate trace file.

Since the PICL library mainly provides functions for interprocess communication, it seemingly cannot be used for the *Sequent Balance*, with its shared memory architecture. That is only true in the sense that there is no machine dependent PICL part developed for this particular hardware type. However, as the goal of this project is visualization, the Sequent Balance programs have to *explicitly* write PICL trace files. Each event in the program is followed by a corresponding line that reports the event type, the current time, and further event specific descriptions according to the PICL trace file format specification.

The trace files produced by PICL and the self-made trace files have to be sorted. PARAGRAPH requires the trace records to contain increasing time stamps. For this reason PARAGRAPH contains a one line shell script, `tracesort`, which calls the UNIX `sort` utility with suitable parameters. While this technique works well for the "real" PICL trace files produced on the Intel Hypercube, it fails for the Sequent Balance trace files. The reason is the low clock resolution, which increases the probability that two or more events will have identical time stamps. Such events might get out of order during sorting and cause PARAGRAPH to display undesired phenomena. Consequently, a short filter program written in C has been developed that takes the unsorted trace file as an input and produces a sorted trace file without disturbing the order of events with identical time stamps.

Chapter 10

IDeC-Specific Displays

PARAGRAPH is extensible in that users can add application-specific displays of their own design. To achieve this, PARAGRAPH contains calls at appropriate points in the source code to routines that provide initialization, data input, event handling, drawing etc. for application-specific displays. For the visualization of parallel IDeC programs, four display types were added. These new displays do not require the trace file to contain special trace records, although they only display data if the trace file contains task definition records (i.e. BLOCK_BEGIN and BLOCK_END).

Mean of Task Run Time Displays show numerical values of the average run time of each task instance. These values are displayed in a two-dimensional matrix in which the individual tasks are on the horizontal axis and the nodes (i.e. processors) are on the vertical axis. Further on there is both an additional column that shows the average run time for all tasks on each node and an additional row that shows the run time mean value for each task on all nodes. Finally there is a single matrix element that displays the average run time of all tasks on all nodes.

Standard Deviation of Task Run Time Displays show numerical values of the standard deviation of each task instance. The visual appearance is the same as in the Mean of Task Run Time displays.

Count of Task Instances Displays show numerical values of the number of times each task was executed. The visual appearance is the same as in the Mean of Task Run Time display.

Histogram Displays show the run time distribution of each task instance. Every time a task instance has terminated, its run time is put in one of twenty[1] classes. Each class is equally wide; the lower bound of the first class and the upper bound of the last class are obtained from the shortest and longest task instance run time, respectively. While the heights of the blocks correspond to the percentage of task instances whose run time is in the respective class, the number displayed above each block shows the absolute number of task instances in each class.

The user can choose which node and which task he wants to observe. Optionally a histogram of all nodes and/or all tasks is possible. The number of the chosen node and the number of the chosen task are displayed at the bottom of the window. With a mouse-click in these two windows, the user is able to change the current node and task number[2]. The choice of the current node/task is also possible in any of the three previously mentioned displays. Clicking on any particular number (mean etc.) in these displays causes the according node and task number to become the current node and task number to be displayed in the histogram.

Below the histogram a small slider provides additional information about the mean value and standard deviation of the chosen histogram. The slider ranges from $\mu - \sigma$ to $\mu + \sigma$, where μ is the mean and σ is the standard deviation. The mean value is depicted by a small vertical line in the middle of this slider.

10.1   Display Categories

The display types of PARAGRAPH which were used in this part of the book can be categorized as follows:

Static Displays (*Utilization Summary,   Concurrency Profile* and all application-specific displays) show certain program states related to the whole program execution.[3]

Gantt Charts (*Utilization Gantt Chart,   Space-Time Diagram , Critical Path Display* and *Task Gantt Chart*) have time on the horizontal axis and the processor number on the vertical axis. The scale of the

---

[1]In the current implementation this number cannot be changed by the user without modifying the program. However, to change the number of classes only one line of an include file has to be modified accordingly. After recompiling the program the new value of the number of classes goes into effect.

[2]The left mouse button decrements the node/task number, the right mouse button increments the node/task button, and the middle button switches to all nodes/tasks.

[3]All of these displays except the *Concurrency Profile* can be animated during the program run. However, this dynamic feature is not easy to illustrate on paper and is therefore not dealt with in this book.

time axis may be changed by the user even during the visualization. If additional space is required, these displays may scroll horizontally. The scale of the figures in Chapter 11 and 12 is chosen in such a way that they either show the entire program run or a certain time interval of special interest.

Dynamic Displays (*Kiviat Diagram, Communication Matrix Display* and *Animation Display*) change their appearance in place dynamically. Therefore the hardcopies reflect only one instance in time. The chosen time instance of the figures demonstrates a typical or interesting behavior.

Chapter 11

Shared Memory IDeC Methods

This chapter is divided into three parts. In each part a specific algorithm for a shared memory architecture (Sequent Balance) is discussed with the aid of different kinds of visualizations. For each algorithm two representative sets of parameters are chosen. The first parameter set, called *optimal*, is chosen in such a way that all processors are utilized as uniformly as possible. The second parameter set, called *degenerated*, tries to push the algorithm to extreme behavior. This is reached in a different way for each parallel IDeC program because it is algorithm dependent.

All programs executed on the Sequent Balance have the following parameters in common: the number of intervals to be calculated is 50, for each interval the basic step and 9 defect correction steps are required, the run time uncertainty for both the basic step and for any defect correction step is 30 percent, and the relation of the number of operations for the basic step and for each defect correction step is 1:1. Parameter sets differ in the following ways: the *optimal* parameter sets of all 3 programs use 10 processors and have a *step size rejection* probability of 10 percent, while the *degenerated* parameter sets use only 6 processors for IDeC$_{order}$ methods and 9 processors for IDeC$_{time}$ methods respectively. The step size rejection probabilities of all *degenerated* parameter sets is generally assumed to be 60 percent each.

11.1   IDeC$_{order}$ Methods

11.1.1   Optimal Parameters

The basic idea of the IDeC$_{order}$ method is best revealed by the *Utilization Gantt Chart* and *Task Gantt Chart* displays (Figures 11.1, 11.2). The *Task Gantt Chart* display shows that processor 0 has to calculate all basic steps (black color), and processor k ($k > 0$) has to calculate all k^{th} defect correction steps (indicated by specific patterns). White regions indicate overhead and idle times. These two different states can be distinguished in the *Utilization Gantt Chart* display, where overhead time is shown solid and idle time is shown blank.

At first sight it seems surprising that there is such a small overhead. But on one hand the definition of overhead time *excludes* the time when a processor is waiting for a message (this is classified as idle time, colored white in both displays). On the other hand the actual overhead (i.e. the time spent in communication routines) is very small on a shared-memory architecture. In fact there is no communication taking place in the usual sense (message passing) on such architectures. Overhead includes only the time needed for the synchronization of the concurrent access to the shared memory with semaphores. For this reason overhead time on the Sequent Balance will not be further discussed.

In both displays *white spots* (i.e. idle times) are apparent. They are caused by step size rejections of the basic step. Since every defect correction step can only start after its prior defect correction step (which could also be the basic step) has terminated, every step size rejection causes a series of white spots, one for each succeeding processor.

Figures 11.3 and 11.4 are zoomed parts of the *Task Gantt Chart* display. They illustrate in more detail the start-up and the closing phase of the execution, where the effects of step size rejections are more clearly visible.

The execution dependencies are even more obvious in the *Space-Time Diagram* (Figures 11.5 shows the entire run, Figure 11.6 displays the first third of the run). Processor k can start with the calculation of the kth defect correction step only after processor $k-1$ has finished the $(k-1)^{st}$ defect correction step. For this reason the start-up phase in these diagrams looks like stairs leading upward. Interrupted horizontal lines, which indicate processor activity, give hints where step size rejections have occurred. The series of idle periods in succeeding processors is, again, easily recognized (white corridors).

The *Utilization Summary* (Figure 11.7) indicates a uniformly balanced load, yet the processors are utilized only to 75 percent. It should be no surprise that the overall utilization of processor 0 is some percentage higher than on the other processors. This is caused by the step size rejections, which force this processor to recalculate the numerical solution on the current interval.

The *Concurrency Profile* (Figure 11.8) indicates that, over 70 percent of the whole run time, at least 8 processors were busy (i.e. doing useful work). Over an additional 10 percent of the run time, 7 processors were busy.

The (snapshot of the) *Communication Matrix Display* (Figure 11.9) clearly points out that each processor k sends information only to processor $k + 1$, except the last processor (no. 9) which does not send anything at all. Similarly each processor k receives signals only from processor $k - 1$, except processor 0, which does not receive any signals at all.

The *Animation Display* (Figure 11.10) shows an instant of time as an example of possible processor states. All processors except no. 2 and no. 4 are working on their respective intervals. Processor 2 has finished its step and has already sent a signal to its successor (no. 3), which in turn has already fetched this signal to start the next defect correction step. However, processor 2 is still idle because it is waiting for a signal from processor 1. Processors 7 and 8 are busy, but the line connecting them indicates that processor 7 has sent data for which processor 8 has not yet asked. In practice this means that after the termination of the current defect correction step (and after sending a signal to processor 9), processor 8 can fetch the signal from processor 7 (thereby clearing the line connecting these two processors) and start with the next interval immediately. Similarly, processor 3 indicates to processor 4 that an interval is ready to be started. Processor 4 is just sending a signal to processor 5, thereafter it will receive the signal from processor 3. Processor 5, however, is still busy with the prior interval.

The next three displays show numerical values[1] concerning the run time of each task instance: the mean (Figure 11.11), the standard deviation (Figure 11.12) and the count, i.e. the number of completed task instances (Figure 11.13). All three displays point out very clearly that all basic steps (task 0) are calculated on processor 0 and all k^{th} defect correction steps are calculated on processor k.

The basic step has the highest mean run time caused by step size rejections. This becomes even more apparent in the display for the standard deviations. The basic step has a far higher standard deviation than any defect correction step. The counts of each task instance demonstrate that each of the 50 intervals is processed on each processor once.

The last three displays present histograms of the task run times. Figure 11.14, a histogram of all tasks, shows that almost all task instances needed between 6 and 12 time units to complete. Only a few instances needed up to 26 time units. These few must have been basic steps, because the displayed numbers on top of the blocks correspond to those of Figure 11.15, the histogram of the basic step. Figure 11.16 shows the equally distributed histogram of a defect correction step.

---

[1]The given numbers should be considered as general *time units*. They do *not* represent milliseconds or other common time units.

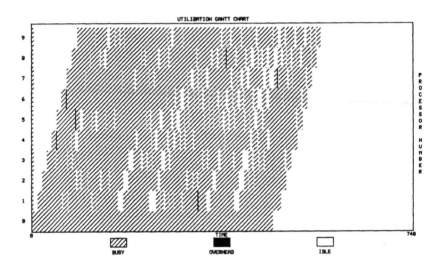

FIGURE 11.1: IDeC$_{order}$ optimal – Utilization Gantt Chart

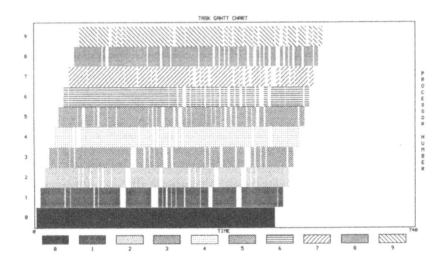

FIGURE 11.2: IDeC$_{order}$ optimal – Task Gantt Chart

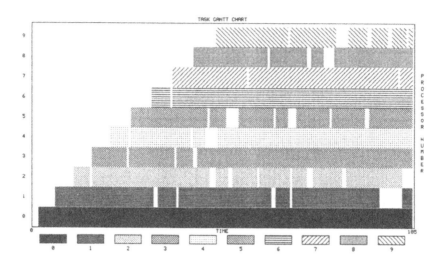

FIGURE 11.3: IDeC_{order} optimal – Task Gantt Chart (start-up)

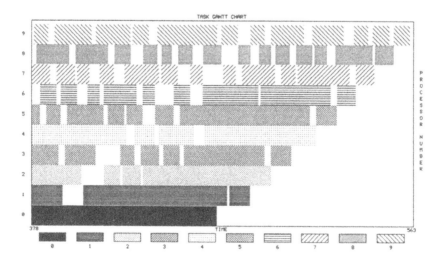

FIGURE 11.4: IDeC_{order} optimal – Task Gantt Chart (close down)

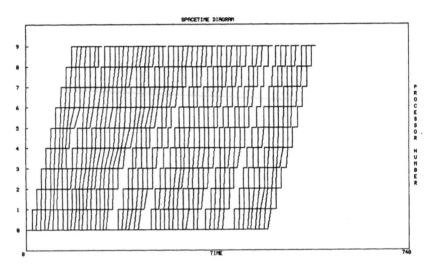

FIGURE 11.5: IDeC$_{order}$ optimal – Space-Time Diagram

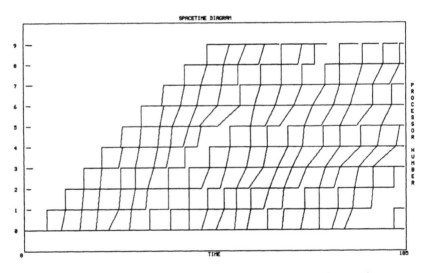

FIGURE 11.6: IDeC$_{order}$ optimal – Space-Time Diagram (start-up)

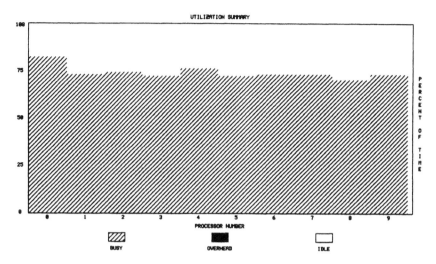

FIGURE 11.7: IDeC_{order} optimal – Utilization Summary

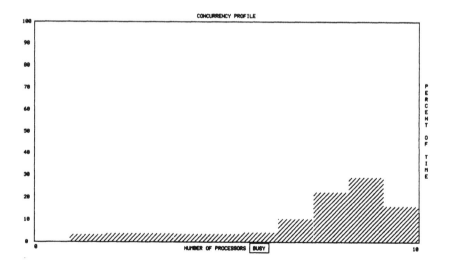

FIGURE 11.8: IDeC_{order} optimal – Concurrency Profile

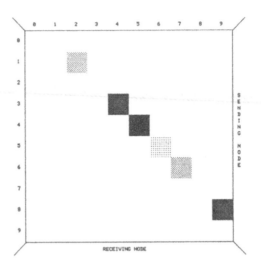

FIGURE 11.9: IDeC$_{order}$ optimal – Communication Matrix (snapshot)

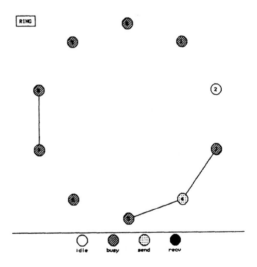

FIGURE 11.10: IDeC$_{order}$ optimal – Animation Display (snapshot)

	task0	task1	task2	task3	task4	task5	task6	task7	task8	task9	total
node0	9										9
node1		8									8
node2			8								8
node3				8							8
node4					9						9
node5						8					8
node6							8				8
node7								8			8
node8									8		8
node9										8	8
total	9	8	8	8	9	8	8	8	8	8	8

FIGURE 11.11: IDeC$_{order}$ optimal – Task Instance Mean Run Time

	task0	task1	task2	task3	task4	task5	task6	task7	task8	task9	total
node0	4.2012										4.2012
node1		1.4148									1.4148
node2			1.5331								1.5331
node3				1.4472							1.4472
node4					1.3775						1.3775
node5						1.4041					1.4041
node6							1.4757				1.4757
node7								1.408			1.408
node8									1.4967		1.4967
node9										1.4318	1.4318
total	4.2012	1.4148	1.5331	1.4472	1.3775	1.4041	1.4757	1.408	1.4967	1.4318	1.9393

FIGURE 11.12: IDeC$_{order}$ optimal – Task Instance Standard Deviation Run Time

	task0	task1	task2	task3	task4	task5	task6	task7	task8	task9	total
node0	50										50
node1		50									50
node2			50								50
node3				50							50
node4					50						50
node5						50					50
node6							50				50
node7								50			50
node8									50		50
node9										50	50
total	50	50	50	50	50	50	50	50	50	50	500

FIGURE 11.13: IDeC$_{order}$ optimal – Task Instance Count

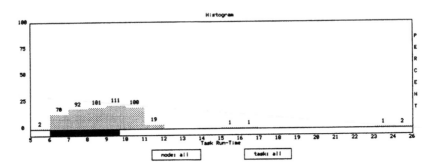

FIGURE 11.14: IDeC$_{order}$ optimal – Task Run Time Histogram (all tasks)

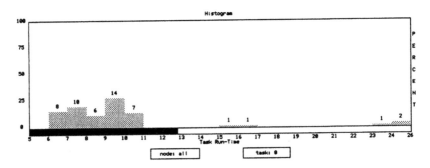

FIGURE 11.15: IDeC$_{order}$ optimal – Task Run Time Histogram (basic step)

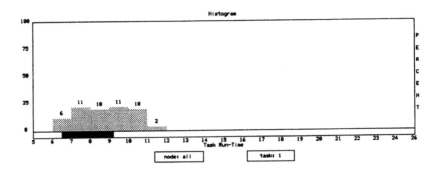

FIGURE 11.16: IDeC$_{order}$ optimal – Task Run Time Histogram (defect correction)

11.1.2   Degenerated Parameters

For the visualization of IDeC$_{\mathrm{order}}$ methods with the *degenerated* parameter set, the processor number was decreased and the probability for step size rejections was increased.

Taking a look at the *Utilization Gantt Chart* (Figure 11.17) it is obvious that processors 0 and 1 have much less work to do than the other processors.[2] Switching attention to the *Task Gantt Chart* display (Figure 11.18), it becomes clear that each of the processors 2 through 5 has to calculate two defect correction steps for each interval, while processor 1 calculates only one defect correction step for each interval and is idle for about half the time. Notice the white spots of processor 1, which are undoubtedly larger than those of the *Task Gantt Chart* display of the *optimal* parameter choice (Figure 11.2). This is due to the higher step size rejection probability of the basic step, which causes processor 0 to need additional time for more intervals than with a lower step size rejection probability.

The *Space-Time Diagram* (Figures 11.19, 11.20) emphasizes the delay of messages sent from processor 1 to processor 2. This delay becomes larger with each interval. The second display points out the consequences of step size rejections in more detail. Especially the basic step of the fifth and sixth interval cause the other processors to become idle for some time. At later points of time the step size rejections do not influence the overall performance because processors 2 to 5 are still busy with prior intervals. Nevertheless, the presence of step size rejections can not be denied when comparing the regular appearance of signals between processors 2 and 3 (also between processors 3 and 4, and between processors 4 and 5) and the irregular occurrence of signals between processors 0 and 1.

The *Utilization Summary* (Figure 11.21) confirms that processors 2 to 5 are utilized more than 80 percent of the time, while processor 1 is idle more than 50 percent. Because of the high step size rejection probability, processor 0 is busy 70 percent of the overall run time.

The *Concurrency Profile* (Figure 11.22) shows that for 90 percent of the time at least four processors are busy. Notice, however, that the states "four processors busy", "five processors busy" and "six processors busy" occur approximately with the same frequency. This is a contrast to the *Concurrency Profile* of the *optimal* parameter set (Figure 11.8).

The mean run time of the basic step is nearly twice that of any defect correction step due to the high number of step size rejections (Figure 11.23). Also the standard deviation of the basic step becomes higher (Figure 11.24) compared to the *optimal* parameter set (Figure 11.12).

---

[2]Note that a different zoom factor was necessary to the display the whole run within a single display. This also applies to other displays of IDeC$_{\mathrm{order}}$ methods with the *degenerated* parameter set, which have time on their horizontal axis.

The histograms reveal that many task instances of the basic step need additional time (Figure 11.27), while the histogram of any defect correction step (Figure 11.28) looks similar to that of the *optimal* parameter set (Figure 11.16).

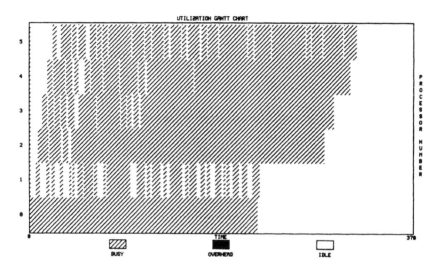

FIGURE 11.17: IDeC~order~ degenerated – Utilization Gantt Chart

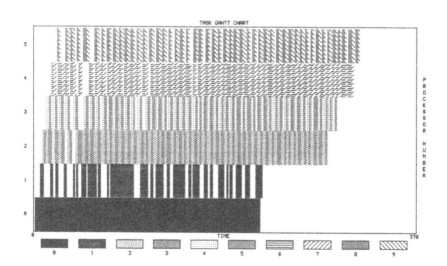

FIGURE 11.18: IDeC~order~ degenerated – Task Gantt Chart

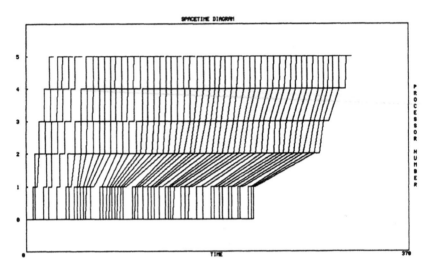

FIGURE 11.19: $IDeC_{order}$ degenerated – Space-Time Diagram

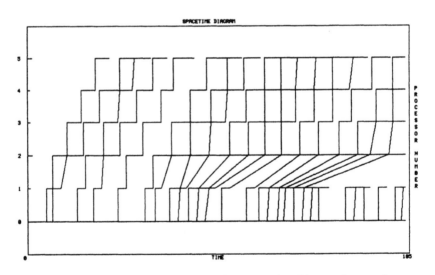

FIGURE 11.20: $IDeC_{order}$ degenerated – Space-Time Diagram (start-up)

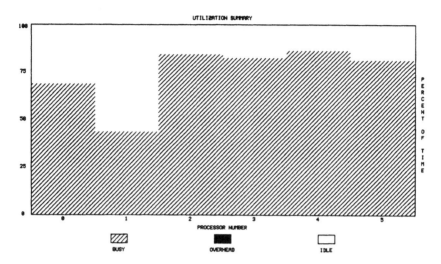

FIGURE 11.21: IDeC_order degenerated – Utilization Summary

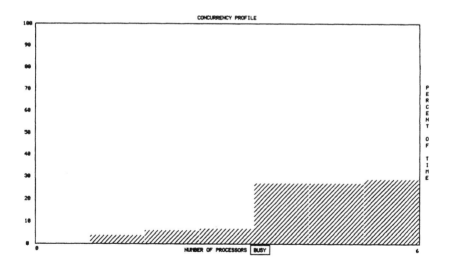

FIGURE 11.22: IDeC_order degenerated – Concurrency Profile

	task0	task1	task2	task3	task4	task5	task6	task7	task8	task9	total
node0	13										13
node1		8									8
node2			8	8							8
node3					8	8					8
node4							8	8			8
node5									8	8	8
total	13	8	8	8	8	8	8	8	8	8	9

FIGURE 11.23: $IDeC_{order}$ degenerated – Task Instance Mean Run Time

	task0	task1	task2	task3	task4	task5	task6	task7	task8	task9	total
node0	6.4695										6.4695
node1		1.4361									1.4361
node2			1.3321	1.6714							1.5118
node3					1.5743	1.5237					1.5574
node4							1.6249	1.5			1.5658
node5									1.3078	1.4479	1.3955
total	6.4695	1.4361	1.3321	1.6714	1.5743	1.5237	1.6249	1.5	1.3078	1.4479	2.9317

FIGURE 11.24: $IDeC_{order}$ degenerated – Task Instance Standard Deviation Run Time

	task0	task1	task2	task3	task4	task5	task6	task7	task8	task9	total
node0	50										50
node1		50									50
node2			50	50							100
node3					50	50					100
node4							50	50			100
node5									50	50	100
total	50	50	50	50	50	50	50	50	50	50	500

FIGURE 11.25: $IDeC_{order}$ degenerated – Task Instance Count

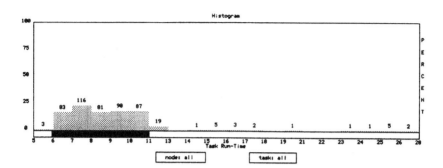

FIGURE 11.26: IDeC_{order} degenerated – Task Run Time Histogram (all tasks)

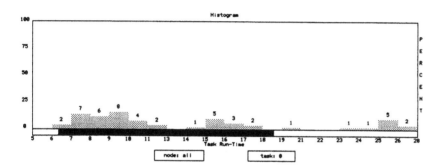

FIGURE 11.27: IDeC_{order} degenerated – Task Run Time Histogram (basic step)

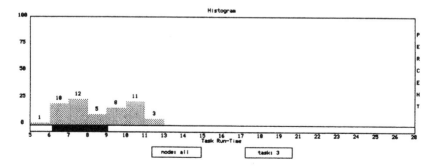

FIGURE 11.28: IDeC_{order} degenerated – Task Run Time Histogram (defect correction)

11.2 IDeC_{time} Methods

11.2.1 Optimal Parameters

The IDeC$_{time}$ *Utilization Gantt Chart* display (Figure 11.29) seems to be very similar to the one of the IDeC$_{order}$ algorithm. (Figure 11.1). Upon closer examination it can be seen that processor 0 also has "white spots" (idle times) over the whole run.

The *Task Gantt Chart* display (Figure 11.30) gives the best demonstration of IDeC$_{time}$ methods. Each processor calculates a whole interval (i.e. the basic step *and* all defect correction steps) before starting another interval. However, after the termination of the basic step of the current interval the following processor can already start with the basic step of the next interval. Similarly, after the termination of the kth defect correction step of a particular interval, the succeeding processor can start with the kth defect correction step of the next interval. Zoomed versions of the *Task Gantt Chart* display (Figures 11.31, 11.32) clarify that the effect of step size rejections is not as drastic as in IDeC$_{order}$ algorithms.

The *Space-Time Diagram* (Figure 11.33) appears confusing. Therefore Figure 11.34 presents this type of display for only the first third of the run. Although this diagram is similar to the IDeC$_{order}$ *Space-Time Diagram*, it soon becomes obvious that there are additional messages from processor 9 to processor 0, which instructs the latter to start with the next basic step or with the next defect correction step of its current interval, respectively.

Utilization Summary (Figure 11.35) and *Concurrency Profile* (Figure 11.36) are nearly identical to the respective displays of IDeC$_{order}$ algorithms (Figures 11.7, 11.8). One minor difference, apparent in the *Utilization Summary*, is the fact that processor 0 is also utilized approximately to the same percentage of time as the other processors. This, of course, is directly influenced by the algorithm that distributes the execution of the basic step among all processors.

The   source and the destination of signals is visible in the *Communication Matrix Display* (Figure 11.37) and the *Animation Display* (Figure 11.38). These displays can be distinguished from those of IDeC$_{order}$ algorithms (Figures 11.9, 11.10) by the additional communication between processors 9 and 0.

The   displays dealing with task instance run times, i.e. *Task Instance Mean Run Time* (Figure 11.39), *Task Instance Standard Deviation Run Time* (Figure 11.40), and *Task Instance Count* (Figure 11.41), differ drastically to those of IDeC$_{order}$ algorithms (Figures 11.11, 11.12, 11.13). Since all processors calculate all tasks for a number of the intervals, the displayed matrices of the three figures do not have any "holes". However, since each task is processed on each processor only five times (see *Task Instance Count* display), some of the numbers in the other two figures appear surprising. The mean run time of the basic step (task 0) is *not* higher than the mean run time of the defect correction steps. Also the standard deviation of the basic step is only slightly higher than the standard deviation of the defect correction steps. Both results are influenced by the low step size rejection probability (10 percent).

The *Task Run Time Histograms* (Figures 11.42, 11.43, 11.44) are more informative. The run times of the defect correction steps range from 5 to 12 time units, while the run times of the basic step are up to 16 time units long.

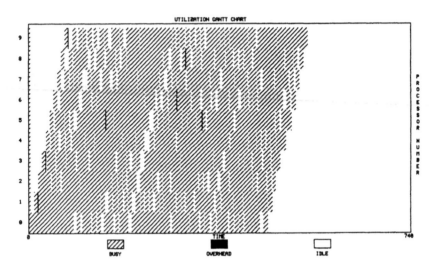

FIGURE 11.29: IDeC$_{time}$ optimal – Utilization Gantt Chart

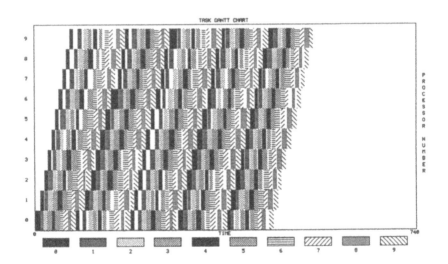

FIGURE 11.30: IDeC$_{time}$ optimal – Task Gantt Chart

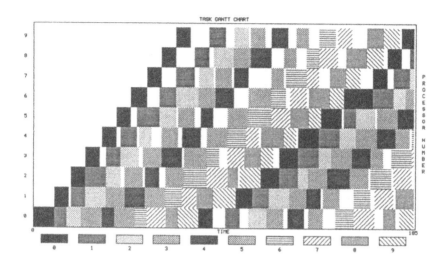

FIGURE 11.31: IDeC$_{time}$ optimal – Task Gantt Chart (start up)

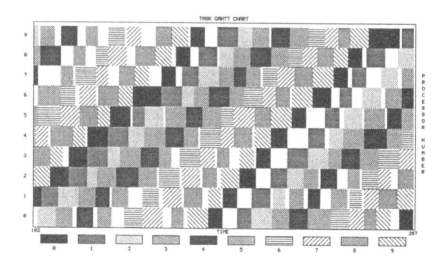

FIGURE 11.32: IDeC$_{time}$ optimal – Task Gantt Chart (near start up)

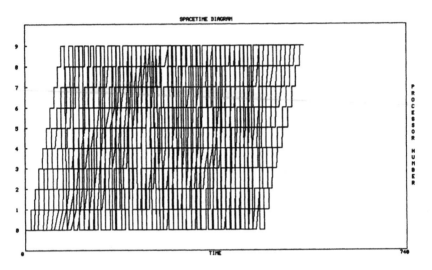

FIGURE 11.33: IDeC$_{time}$ optimal – Space-Time Diagram

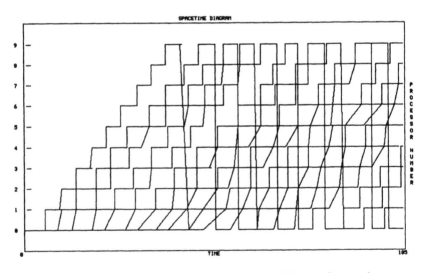

FIGURE 11.34: IDeC$_{time}$ optimal – Space-Time Diagram (start up)

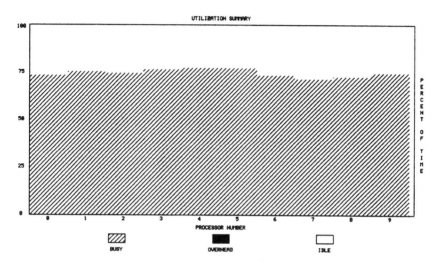

FIGURE 11.35: IDeC$_{time}$ optimal – Utilization Summary

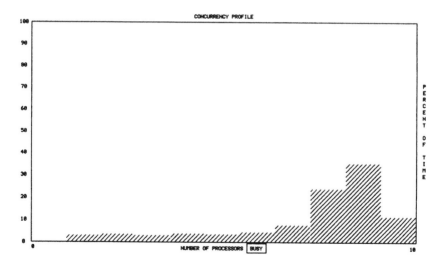

FIGURE 11.36: IDeC$_{time}$ optimal – Concurrency Profile

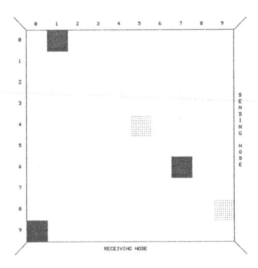

FIGURE 11.37: IDeC$_{\text{time}}$ optimal – Communication Matrix (snapshot)

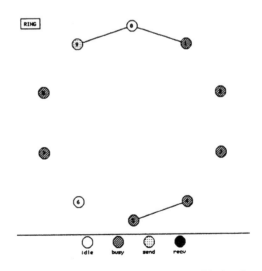

FIGURE 11.38: IDeC$_{\text{time}}$ optimal – Animation Display (snapshot)

	task0	task1	task2	task3	task4	task5	task6	task7	task8	task9	total
node0	8	7	8	8	8	9	8	9	7	9	8
node1	8	7	9	8	8	9	9	8	9	8	8
node2	8	9	7	8	8	7	9	8	7	8	8
node3	8	9	8	8	9	9	8	9	8	8	8
node4	8	9	8	9	9	7	8	8	10	8	8
node5	9	7	9	9	7	9	8	8	9	8	8
node6	9	8	8	8	8	9	9	7	8	8	8
node7	7	8	8	8	7	7	10	7	8	8	8
node8	7	9	9	8	8	7	7	8	8	8	8
node9	9	8	7	8	8	8	8	7	8	9	8
total	8	8	8	8	8	8	8	8	8	8	8

FIGURE 11.39: IDeC$_{time}$ optimal – Task Instance Mean Run Time

	task0	task1	task2	task3	task4	task5	task6	task7	task8	task9	total
node0	1.4697	1.0954	1.0198	.9798	1.3266	1.6248	1.2	1.1662	.8	1.0954	1.407
node1	.74833	1.1662	1.6	.74833	1.6	.74833	1.7436	.8	.9798	1.7205	1.3898
node2	.8	.74833	1.6248	.8	1.4697	1.3565	1.2	1.6248	1.3565	1.6248	1.4562
node3	1.4697	1.1662	1.4142	1.6733	.63246	1.6733	1.833	1.2	1.6	1.6	1.5395
node4	1.3565	.9798	1.3565	.9798	1.0954	1.3565	1.6248	1.833	.63246	1.1662	1.4865
node5	1.3565	1.8547	.74833	1.7889	.89443	.4899	1.7205	1.2	1.1662	1.6	1.5519
node6	2.7857	1.3266	1.9391	.74833	1.4142	1.8547	.9798	.8	1.6733	1.3565	1.7136
node7	.4	1.1662	1.1662	.4899	1.3565	1.3565	1.4967	1.2	1.4142	1.4697	1.4051
node8	.4	1.0954	1.2649	1.0954	1.833	1.0198	.4	.63246	1.0954	.4	1.2207
node9	3.2496	1.3266	.74833	.89443	.8	1.3266	1.0954	.74833	1.6	1.4142	1.6228
total	1.7884	1.4091	1.4832	1.2	1.4232	1.527	1.6055	1.3711	1.5819	1.4637	1.4992

FIGURE 11.40: IDeC$_{time}$ optimal – Task Instance Standard Deviation Run Time

	task0	task1	task2	task3	task4	task5	task6	task7	task8	task9	total
node0	5	5	5	5	5	5	5	5	5	5	50
node1	5	5	5	5	5	5	5	5	5	5	50
node2	5	5	5	5	5	5	5	5	5	5	50
node3	5	5	5	5	5	5	5	5	5	5	50
node4	5	5	5	5	5	5	5	5	5	5	50
node5	5	5	5	5	5	5	5	5	5	5	50
node6	5	5	5	5	5	5	5	5	5	5	50
node7	5	5	5	5	5	5	5	5	5	5	50
node8	5	5	5	5	5	5	5	5	5	5	50
node9	5	5	5	5	5	5	5	5	5	5	50
total	50	50	50	50	50	50	50	50	50	50	500

FIGURE 11.41: IDeC$_{time}$ optimal – Task Instance Count

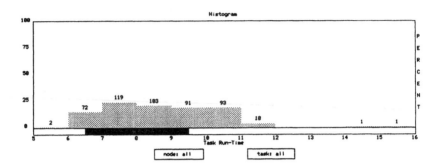

FIGURE 11.42: IDeC$_{\text{time}}$ optimal – Task Run Time Histogram (all tasks)

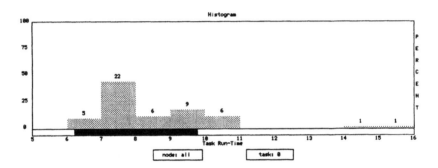

FIGURE 11.43: IDeC$_{\text{time}}$ optimal – Task Run Time Histogram (basic step)

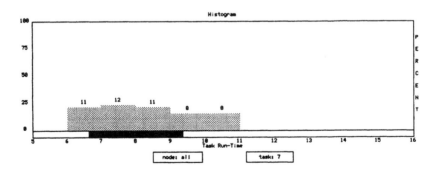

FIGURE 11.44: IDeC$_{\text{time}}$ optimal – Task Run Time Histogram (defect correction)

11.2.2   Degenerated Parameters

For the visualization of $IDeC_{time}$ methods with the *degenerated* parameter set, the processor number was decreased and the probability for step size rejections was increased.

The *Utilization Gantt Chart* display (Figure 11.45) already indicates a poor performance. Many parts of the whole display are white (idle times). The cause of this is visible in the *Task Gantt Chart* display (Figure 11.46). Many step size rejections (wide black bars) cause the start of successive intervals to be delayed. Another aspect apparent in the. *Task Gantt Chart* display can be seen on the far left, near the closing phase. Since the number of processors is not a divisor of the number of total intervals, some of the processors are idle for a long time at the end of the execution, while the others calculate the remaining intervals.

The long periods of idle time can also be seen in the *Space-Time Diagram* (Figures 11.47, 11.48, 11.49). The first display, showing the whole execution, may be too confusing. Therefore two additional figures zoom parts of the whole display to emphasize details. Any two neighboring communication lines[3] which are *not* approximately parallel to each other indicate a longer run time (step size rejection) of the task that is executed between the receipt of these two messages.

The *Kiviat Diagram* (Figure 11.50) displays one snapshot at a time near the far right in Figure 11.48. Processors 3 and 4 are not fully utilized, while the other processors are.[4] The dynamic appearance of this display is noteworthy: for the entire time a few processors are not fully utilized. Since the set of partly utilized processors changes all the time, the whole diagram looks like a turning disc from which a piece has been cut away.

The *Utilization Summary* (Figure 11.51) indicates a poor overall performance. The fact that processors 0 to 4 have to calculate one additional interval at the end of the execution is also visible.

The *Concurrency Profile* (Figure 11.52) signals that all nine processors are working concurrently only for a small portion of the whole run time. Most of the time (about 75 percent) 5 to 8 processors are busy.

The *Animation Display* (Figure 11.53) reveals two aspects: there are successive nodes which are idle (colored white), and there are successive nodes whose message queues contain signals for which the respective node has not yet asked (connecting lines). In other words, some of the processors do not have any calculations to work on while others are overloaded.

---

[3]All lines except the horizontal ones are communication lines; horizontal lines indicate processor activity.

[4]This type of diagram does not show the *current* processor utilization (this would be either 0 or 1 for each processor), but the processor utilization during a small fixed interval.

Both the mean run time and the standard deviation of the basic steps are higher than those of all defect correction steps (Figures 11.54, 11.55). This fact is also visible in the *Task Run Time Histograms* (Figures 11.56, 11.57).

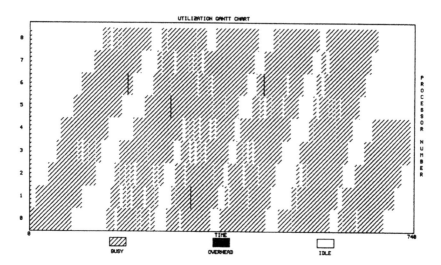

FIGURE 11.45: IDeC$_{time}$ degenerated – Utilization Gantt Chart

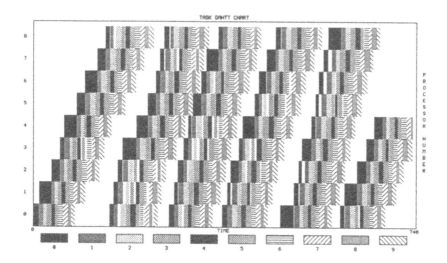

FIGURE 11.46: IDeC$_{time}$ degenerated – Task Gantt Chart

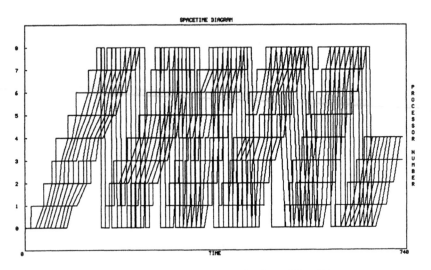

FIGURE 11.47: IDeC$_{time}$ degenerated – Space-Time Diagram

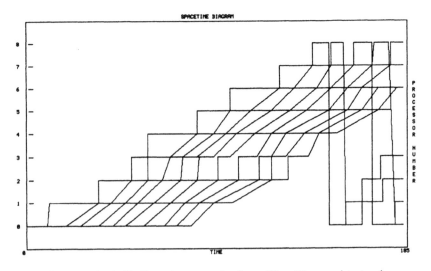

FIGURE 11.48: IDeC$_{time}$ degenerated – Space-Time Diagram (start up)

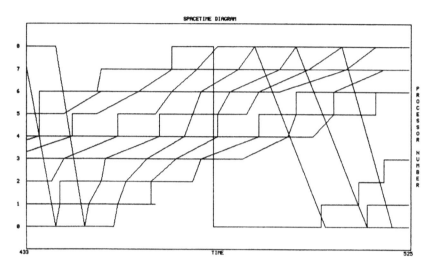

FIGURE 11.49: IDeC$_{time}$ degenerated – Space-Time Diagram (near close down)

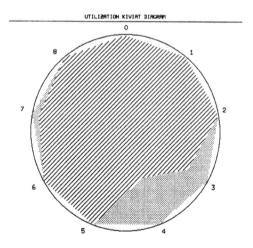

FIGURE 11.50: IDeC$_{time}$ degenerated – Kiviat Diagram

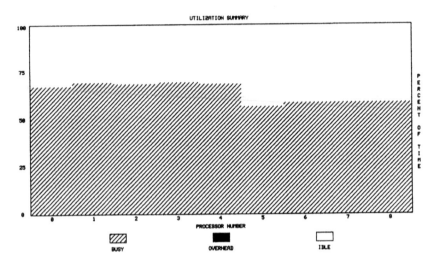

FIGURE 11.51: IDeC$_{time}$ degenerated – Utilization Summary

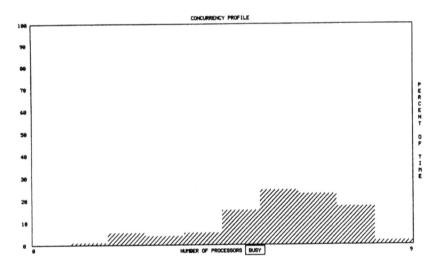

FIGURE 11.52: IDeC$_{time}$ degenerated – Concurrency Profile

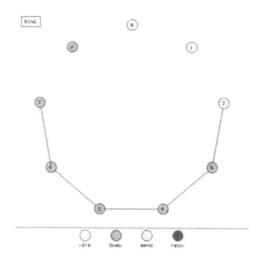

FIGURE 11.53: IDeC$_{time}$ degenerated – Animation Display (snapshot)

	task0	task1	task2	task3	task4	task5	task6	task7	task8	task9	total
node0	12	9	8	8	8	8	8	9	7	7	9
node1	13	9	8	8	8	8	8	8	8	8	9
node2	13	8	8	8	9	9	9	8	7	9	9
node3	14	8	7	7	8	9	9	7	8	9	9
node4	17	8	8	8	8	7	8	9	8	8	9
node5	12	9	8	8	8	7	7	8	9	8	9
node6	13	9	8	8	8	8	8	8	10	9	9
node7	15	8	9	9	8	9	8	7	8	8	9
node8	12	8	10	10	8	8	9	7	8	8	9
total	13	8	8	8	8	8	8	8	8	8	9

FIGURE 11.54: IDeC$_{time}$ degenerated – Task Instance Mean Run Time

	task0	task1	task2	task3	task4	task5	task6	task7	task8	task9	total
node0	6.3246	.8165	1.7951	.94281	1.675	1.7951	.89753	1	1.3437	.4714	2.6672
node1	7.8316	1.4907	1.6997	1.7951	1.2472	.4714	1.2472	1.5986	1.2134	.68718	3.171
node2	4.3076	1.2472	1.6997	1.8856	.76376	1.2134	1.4142	1.5	1.4625	1.3844	2.4296
node3	7.0238	1.8634	1.6997	1	1.5275	1.4142	1	1.6997	1.5723	1.0672	3.2497
node4	7.2053	.94281	1.3437	2.0344	1.4907	.94281	1.4625	1.1547	.76376	1.3437	3.6854
node5	4.4542	1.6	1.3266	1.1662	1.6733	1.9596	1.0198	.89443	.74833	1.6	2.3672
node6	5.8652	1.3565	1.7205	1.6	1.3266	1.2	1.7205	.8	1.0198	1.3266	2.7229
node7	5.5353	1.3266	1.7436	1.4697	1.2649	.74833	1.4697	1.1662	1.8974	1.4142	2.989
node8	6.0663	1.4697	.4899	1.4697	1.4142	1.6	.63246	1.1662	1.0954	1.3565	2.6348
total	6.3777	1.4832	1.6733	1.6919	1.4211	1.5658	1.3629	1.5021	1.5032	1.3454	2.9272

FIGURE 11.55: IDeC$_{time}$ degenerated – Task Instance Standard Deviation Run Time

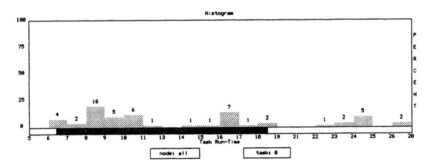

FIGURE 11.56: IDeC$_{time}$ degenerated – Task Run Time Histogram (basic step)

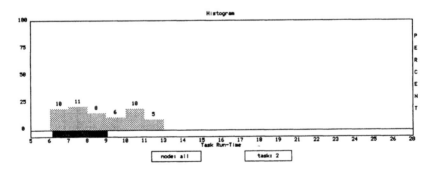

FIGURE 11.57: IDeC$_{time}$ degenerated – Task Run Time Histogram (defect correction)

11.3 IDeC$_{\text{adaptive}}$ Methods

11.3.1 Optimal Parameters

The IDeC$_{\text{adaptive}}$ *Utilization Gantt Chart* display (Figure 11.58) does not seem to reveal any algorithm specific characteristics. And, indeed, the *Task Gantt Chart* display (Figure 11.59) is, in the first place, also not as informative as the corresponding displays of the previously discussed algorithms. This is due to the fact that the order of calculations is not statically scheduled. Rather, each processor that has finished its current calculations looks for an interval whose basic step or whose defect correction step could already be started. This concept yields an excellent load balance, since the additionally required communication overhead is rather cheap on shared memory architectures. This becomes more obvious in zoomed parts of the *Task Gantt Chart* display (Figures 11.60, 11.61). Even step size rejections do not create long idle periods on the other processors because those other processors may continue with "older" intervals that have not been processed yet.

As a result of the dynamic load balancing, the number of messages has increased (doubled). Accordingly the *Space-Time Diagram* (Figure 11.62) appears very crowded. Figure 11.63 displays approximately the first 20 percent of the whole diagram to make the basic idea of the algorithm visible. After processor 0 has finished calculating the basic step of the first interval it sends two signals: the first to initiate the basic step of the second interval (processor 1), and the second to initiate the first defect correction step of the first interval (processor 6). After processor 1 has finished with the basic step of the second interval, it sends one signal to processor 2, which starts to calculate the basic step of the third interval, and another signal to processor 7. Processor 7, however, may not start with the first defect correction step of the second interval. It has to wait for the termination of the basic step of the second interval, which is calculated on processor 6. In the same way, all other tasks are started: all basic steps and all defect correction steps of the first interval need only one signal to be initiated. All other defect correction steps, however, have to wait for the receipt of two signals before they may be started: one signal from the preceding defect correction step and another one from the preceding interval.

Although the *Concurrency Profile* (Figure 11.65) does not differ much from the corresponding displays of the previously discussed algorithms (70 percent of the time 8 or more processors are busy), it should be pointed out that the highest peak of this display indicates that during 35 percent of the time exactly 9 processors have been busy. This is an indicator of good overall performance which is also visible in the *Utilization Summary* (Figure 11.64). Each processor is utilized to approximately 85 percent, which is quite a lot more than with other algorithms.

Communication is possible between any pair of processors (*Communication Matrix Display*, Figure 11.66).

The *Animation Display* (Figure 11.67) is a snapshot after processor 0 has finished its current defect correction step and is just sending two signals to processor 5 and processor 8, respectively, to initiate the calculation of another defect correction step on these processors.

The mean, the standard deviation, and the count of the task instances are shown in Figures 11.68, 11.69, and 11.70, respectively. Task 0, the basic step, has a higher mean and higher standard deviation as usual. Also the *Task Instance Count* display, which lists how many times each task was executed on each individual processor, may be of special interest. Due to the low overall count these numbers differ greatly. Some tasks seem to "prefer" a special processor, while others are more or less equally distributed.

The *Task Run Time Histograms* (Figures 11.71, 11.72, 11.73) present the usual picture. While the run times of any defect correction step are equally distributed with an average of 8 time units, the histogram of the basic step displays some task instances that need multiples of the time of other task instances.

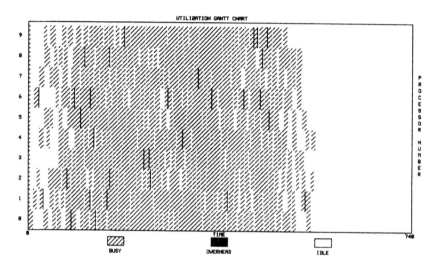

FIGURE 11.58: IDeC_adaptive optimal – Utilization Gantt Chart

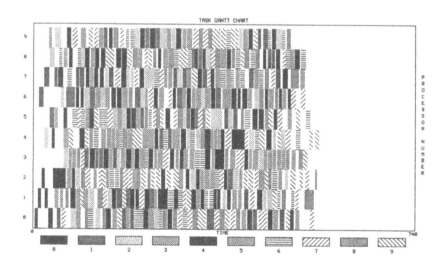

FIGURE 11.59: IDeC_adaptive optimal – Task Gantt Chart

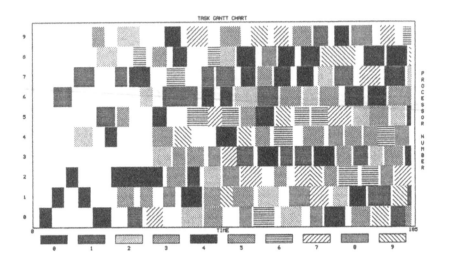

FIGURE 11.60: IDeC$_{adaptive}$ optimal – Task Gantt Chart (start up)

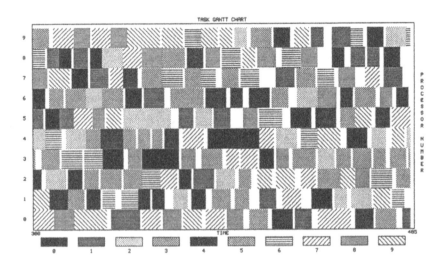

FIGURE 11.61: IDeC$_{adaptive}$ optimal – Task Gantt Chart (mid part)

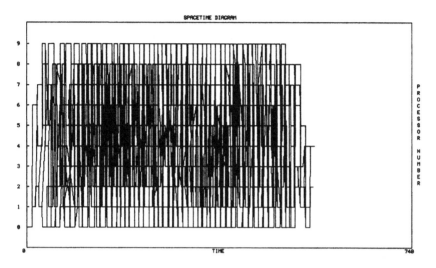

FIGURE 11.62: IDeC_{adaptive} optimal – Space-Time Diagram

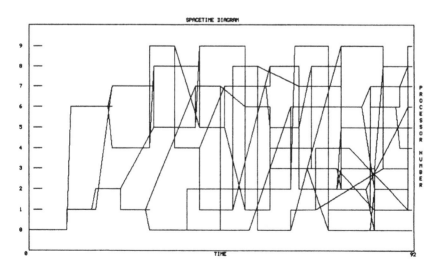

FIGURE 11.63: IDeC_{adaptive} optimal – Space-Time Diagram (start up)

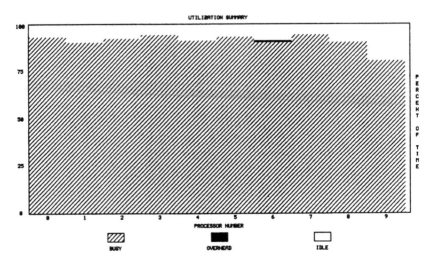

FIGURE 11.64: IDeC$_{\text{adaptive}}$ optimal – Utilization Summary

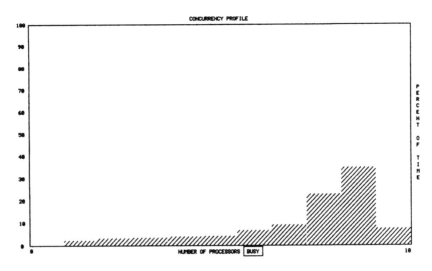

FIGURE 11.65: IDeC$_{\text{adaptive}}$ optimal – Concurrency Profile

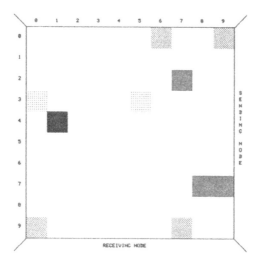

FIGURE 11.66: IDeC_{adaptive} optimal – Communication Matrix (snapshot)

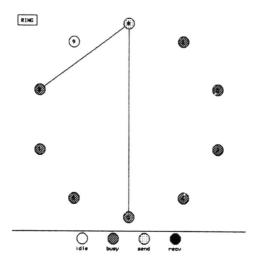

FIGURE 11.67: IDeC_{adaptive} optimal – Animation Display (snapshot)

	task0	task1	task2	task3	task4	task5	task6	task7	task8	task9	total
node0	8	7	10	9	9	9	9	9	8	8	8
node1	9	9	8	7	7	9	8	8	8	8	8
node2	12	10	8	9	7	8	9	8	6	8	8
node3	11	7	8	7	9	8	8	8	7	9	8
node4	12	7	9	8	8	7	9	9	7	8	8
node5	8	8	8	7	9	8	8	8	7	7	8
node6	7	10	8	9	7	7		9	7		8
node7	8	8	8	7	9	9	9	8	8	10	8
node8	8	7	9	8	8	9	8	7	8	8	8
node9	9	8	8	8		10	7	9	9	8	8
total	9	8	8	8	8	8	8	8	8	8	8

FIGURE 11.68: IDeC$_{adaptive}$ optimal – Task Instance Mean Run Time

	task0	task1	task2	task3	task4	task5	task6	task7	task8	task9	total
node0	1.4142	.90351	.4714	.4714	1.0954	1.2247	1.1662	.70711	1.3397	.74833	1.2329
node1	3.7351	1.2247	1.6733	1.1662	1.0672	.94281	1.6073	1.2649	1.6	1.118	2.1108
node2	8.9567	0	1.479	2.0616	.70711	1.5119	.70711	1.4569	.94281	1.4659	2.8578
node3	3.448	.82916	1.2472	.4714	1.118	1.5806	1.6997	1.479	.4899	0	2.0397
node4	6.4062	0	1.2583	1.5535	2.1602	.43301	.5	1.4142	0	1.6583	2.7857
node5	1.2472	.76376	1.1606	1.0897	1.118	0	1.5908	1.8257	1.4697	.74833	1.3534
node6	1.2	1.0302	1.6997	1.3748	1.7069	1.6997		1.5	1.2247		1.7168
node7	0	1.3266	1.5	1.2472	1.2778	.8	1.0672	1.3565	1.2247	0	1.3398
node8	1.4983	.5	1.4142	0	1.4907	1.6997	1.536	1.4142	1.4948	1.5	1.5843
node9	1.0498	1.1662	1.5908	1.6		.5	.89443	1.3565	.86603	1.1249	1.347
total	4.061	1.3927	1.527	1.581	1.5154	1.4566	1.4183	1.4595	1.4455	1.3666	1.9365

FIGURE 11.69: IDeC$_{adaptive}$ optimal – Task Instance Standard Deviation Run Time

	task0	task1	task2	task3	task4	task5	task6	task7	task8	task9	total
node0	3	7	3	3	5	4	5	8	7	5	50
node1	9	4	5	5	6	3	6	5	5	4	52
node2	3	1	4	4	4	7	4	7	6	11	51
node3	6	4	3	3	4	17	3	4	5	1	50
node4	5	2	6	11	3	4	6	3	2	8	50
node5	3	6	7	4	6	2	7	6	5	5	51
node6	5	7	6	10	9	3		4	4		48
node7	2	10	6	3	7	5	6	5	4	1	49
node8	7	4	3	2	6	3	8	3	8	8	52
node9	7	5	7	5		2	5	5	4	7	47
total	50	50	50	50	50	50	50	50	50	50	500

FIGURE 11.70: IDeC$_{adaptive}$ optimal – Task Instance Count

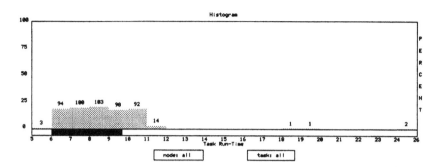

FIGURE 11.71: IDeC_{adaptive} optimal – Task Run Time Histogram (all tasks)

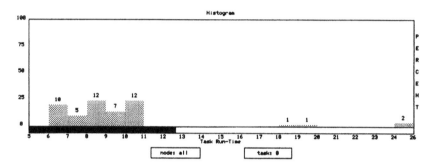

FIGURE 11.72: IDeC_{adaptive} optimal – Task Run Time Histogram (basic step)

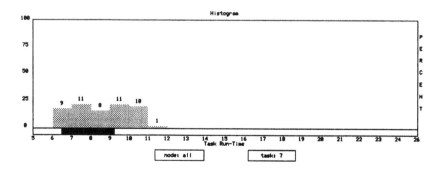

FIGURE 11.73: IDeC_{adaptive} optimal – Task Run Time Histogram (defect correction)

11.3.2 Degenerated Parameters

For the visualization of IDeC$_{adaptive}$ methods with the *degenerated* parameter set, the probability for step size rejections was increased.

The *Utilization Gantt Chart* display (Figure 11.74) of IDeC$_{adaptive}$ methods with the *degenerated* parameter set shows, of course, more idle periods than in the respective display of the *optimal* parameter set. That is not surprising. The question is to what extent the higher step size rejection probability influences the performance. Apart from the very start and the very end each processor seems to be utilized to a quite high rate during the whole run. Also the *Task Gantt Chart* display (Figure 11.75) indicates that any step size rejection does *not* cause series of idle times occurring on other processors.

The *Critical Path Display* (Figure 11.77) is a variation of the *Space-Time Diagram* with the longest serial thread emphasized. If this display is compared with a zoomed version of the *Task Gantt Chart* display (Figure 11.76, both figures are scaled in the same way), it becomes apparent that the sequence of the basic step plus all defect correction steps of the last interval are identical to that longest serial thread. Since any possible IDeC algorithm must last at least that long, it turns out that IDeC$_{adaptive}$ methods are a better choice than IDeC$_{order}$ methods or IDeC$_{time}$ methods, especially when conditions are not optimal.

The good overall performance is also visualized with the help of the *Utilization Summary* display (Figure 11.78). All processors are utilized up to 80 or more percent.

The *Concurrency Profile* (Figure 11.79) highlights another interesting aspect. Almost never are all processors working concurrently. Most of the time (75 percent) 5 to 8 processors are busy. The reason for this is the increased overhead time each processor needs for determining which interval should be calculated next.

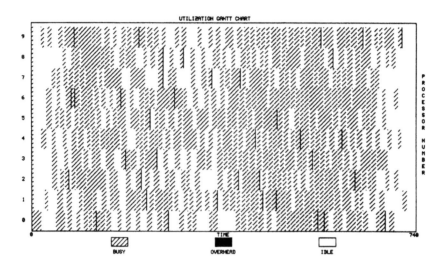

FIGURE 11.74: IDeC~adaptive~ degenerated – Utilization Gantt Chart

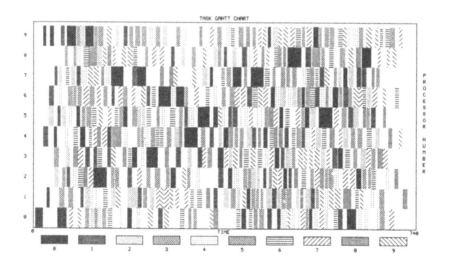

FIGURE 11.75: IDeC~adaptive~ degenerated – Task Gantt Chart

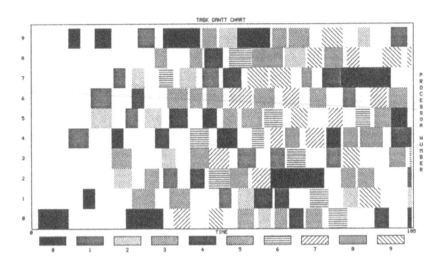

FIGURE 11.76: IDeC_adaptive degenerated – Task Gantt Chart

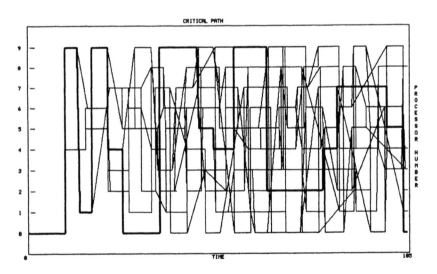

FIGURE 11.77: IDeC_adaptive degenerated – Critical Path

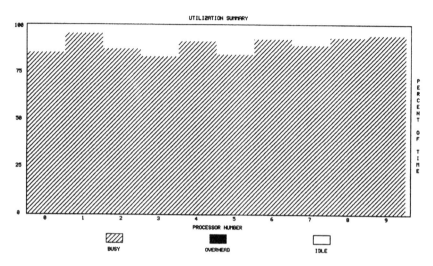

FIGURE 11.78: IDeC_{adaptive} degenerated – Utilization Summary

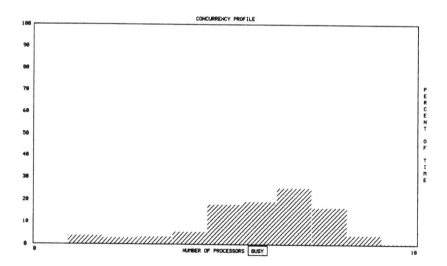

FIGURE 11.79: IDeC_{adaptive} degenerated – Concurrency Profile

Chapter 12

Distributed Memory IDeC Methods

In each part of this chapter a specific algorithm for a distributed memory architecture (Intel Hypercube) is discussed with the aid of various kinds of visualizations. Again, for each algorithm two representative parameter sets are chosen. The *optimal* parameter set tries to utilize all processors as uniformly as possible, and the *degenerated* parameter set tries to push the algorithm to extreme behavior.

All programs executed on the Intel Hypercube have the following common parameters: the number of processors used is 8, the run time uncertainty for both the basic step and for any defect correction step is 30 percent, and the relation of the number of operations for the basic step and for each defect correction step is 1:1. Parameter sets differ in the following ways: the number of intervals to be calculated is 48 except for the *degenerated* parameter set of IDeC$_{time}$ methods where 52 intervals are demanded. The number of defect correction steps required for each interval is 7 except for the *degenerated* parameter set of IDeC$_{order}$ methods where 10 defect correction steps are performed. The *optimal* parameter sets of all programs have a step size rejection probability of 10 percent, while the step size rejection probabilities of the *degenerated* parameter sets are assumed to be 60 percent.

12.1   IDeC$_{\text{order}}$ Methods

12.1.1   Optimal Parameters

The greatest difference between the *Utilization Gantt Chart* (Figure 12.1) and the corresponding display of the Sequent Balance is the increased communication overhead (colored black).[1] This additional overhead is caused by the lack of a shared memory. To exchange information, messages that contain all the necessary data have to be passed from the sending to the receiving node. Thus any communication is much more time consuming than on the Sequent Balance.

In the *Utilization Gantt Chart* "white spots" (idle times) occur only after step size rejections of the basic step. Since the probability of step size rejections is set to only 10 percent, idle times appear almost only at the start-up and the closing phase of the whole run.

The *Task Gantt Chart* display (Figure 12.2) helps in understanding the principle of IDeC$_{\text{order}}$ methods. Each task (i.e. either basic step or defect correction step) is assigned to one particular processor.

The *Space-Time Diagram* (Figure 12.3) reveals that the number of messages is twice as high as that of the same method on the Sequent Balance. On that computer it was only necessary to signal the preceding node to start working on the next defect correction step, while the actual data was passed through the shared memory. On the hypercube, however, all the data necessary for the calculations has to be passed via messages. Therefore messages contain whole matrices and their lengths may be up to some hundred or few thousand bytes. Two messages are required for the defect correction step of each interval. The first contains the solution matrix of the basic step, the other contains the solution matrix of the prior defect correction step.

The order of in- and outgoing messages may be viewed in Figure 12.4, a zoomed version of the start-up of the *Space-Time Diagram*. After the termination of the basic step in the first interval, processor 0 sends the solution matrix of that step to processor 1 and continues with the second interval. After each interval it sends the solution matrix of the basic step of the interval it has just finished calculating to processor 1.

Processor 1 has to wait for the receipt of a message from processor 0 containing the solution matrix of the basic step of the current interval. Thereafter it passes this message on to processor 2 and starts calculating the first defect correction step of the current interval. Finally it sends the solution matrix

---

[1]Note that the *Utilization Gantt Chart* shown suggests that there are overhead times only after *some* defect correction steps. This is *not* true. Indeed, there is a period of overhead after *each* defect correction step. However, the diagram is only capable of revealing longer overhead periods, while the shorter ones disappear for lack of greater printing resolution.

of that defect correction step to processor 2 before continuing with the next interval.

The other processors have to handle even one more event per interval. The kth processor ($2 \leq k \leq 6$) has to wait for two messages before it can start a new interval. Only after the receipt of both the solution matrix of the basic step and the solution matrix of the $(k-1)^{st}$ defect correction step processor k may start its calculations. However, before actually starting, it sends the just received solution matrix of the basic step to its succeeding node. After finishing the defect correction step in the current interval, processor k also sends the solution matrix of the kth defect correction step to processor $k+1$.

The last processor (no. 7 in Figure 12.4) only receives messages and does not send messages anywhere.

The *Utilization Summary* (Figure 12.5) clearly shows the overhead (colored black). Each processor is utilized between 70 and 75 percent, which is slightly worse than on the Sequent Balance. For 50 percent of the time 7 or 8 processors are busy; for 80 percent of the time at least 5 processors are doing useful work (*Concurrency Profile*, Figure 12.6).

The *Communication Matrix Display* (Figure 12.7) emphasizes that each processor k ($0 \leq k \leq 6$) is sending messages to processor $k+1$. Since this diagram only displays *pending messages* (i.e. messages that have already been sent, but that have not yet arrived or for that the receiving node has not yet asked), the number of visible messages may be surprising. But this is caused by the fact that each node passes the information concerning the solution matrix of the basic step to its succeeding node immediately after it has received it. Furthermore, that succeeding node does not need this information before finishing its current calculations. Therefore there is almost always a message at each processor's receiving queue that has to wait to be processed. These pending messages are also easy to spot in the *Space-Time Diagram* (Figure 12.3); they are the ones that "lean" to the right side.

These obtained observations are visible in the *Animation Display* (Figure 12.8), too. There are pending messages between each successive node pair. All processors are busy except processors 4 and 5, which are communicating.

The *Task Instance Mean Run Time* display (Figure 12.9) and the *Task Instance Standard Deviation Run Time* display (Figure 12.10) point out the higher average run time and the higher standard deviation, respectively, of the basic step.[2]

Finally three *Task Run Time Histograms* (Figures 12.12, 12.13, 12.14) graphically present the distribution of the run time of task instances (i) of all tasks, (ii) of the basic step, and (iii) of one defect correction step.

---

[2]The displayed numbers may be assumed as microseconds.

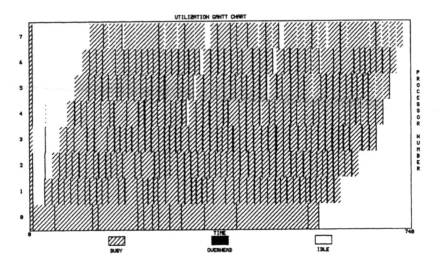

FIGURE 12.1: IDeC$_{order}$ optimal – Utilization Gantt Chart

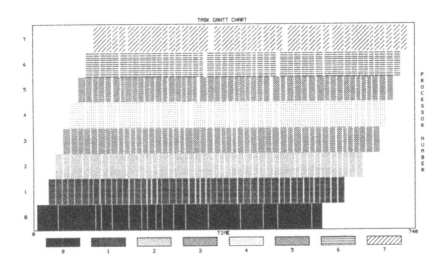

FIGURE 12.2: IDeC$_{order}$ optimal – Task Gantt Chart

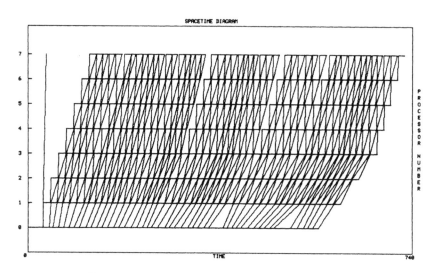

FIGURE 12.3: IDeC$_{order}$ optimal – Space-Time Diagram

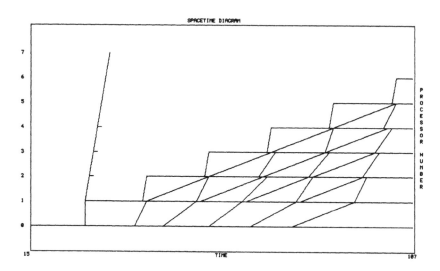

FIGURE 12.4: IDeC$_{order}$ optimal – Space-Time Diagram (start up)

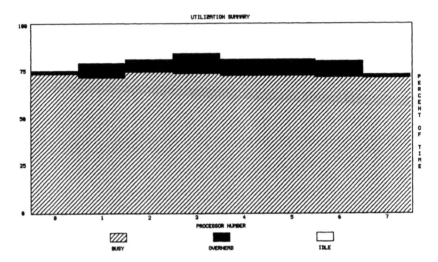

FIGURE 12.5: IDeC$_{order}$ optimal – Utilization Summary

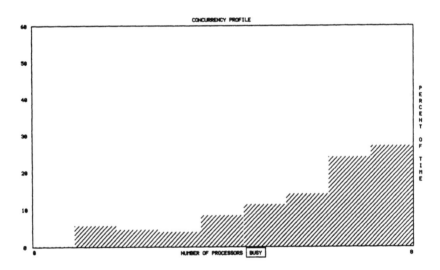

FIGURE 12.6: IDeC$_{order}$ optimal – Concurrency Profile

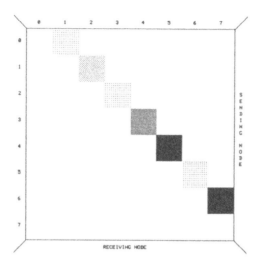

FIGURE 12.7: IDeC$_{order}$ optimal – Communication Matrix (snapshot)

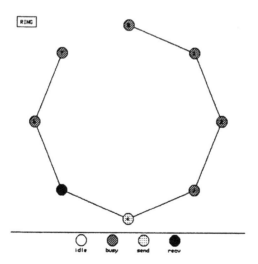

FIGURE 12.8: IDeC$_{order}$ optimal – Animation Display (snapshot)

	task0	task1	task2	task3	task4	task5	task6	task7	total
node0	7226								7226
node1		6912							6912
node2			7060						7060
node3				7082					7082
node4					6926				6926
node5						7003			7003
node6							6975		6975
node7								6941	6941
total	7226	6912	7060	7082	6926	7003	6975	6941	7016

FIGURE 12.9: IDeC$_{order}$ optimal – Task Instance Mean Run Time

	task0	task1	task2	task3	task4	task5	task6	task7	total
node0	2900.8								2900.8
node1		1127.1							1127.1
node2			1131.5						1131.5
node3				1144.3					1144.3
node4					1188				1188
node5						1114.2			1114.2
node6							1132.7		1132.7
node7								1198.4	1198.4
total	2900.8	1127.1	1131.5	1144.3	1188	1114.2	1132.7	1198.4	1488.4

FIGURE 12.10: IDeC$_{order}$ optimal – Task Instance Standard Deviation Run Time

	task0	task1	task2	task3	task4	task5	task6	task7	total
node0	48								48
node1		48							48
node2			48						48
node3				48					48
node4					48				48
node5						48			48
node6							48		48
node7								48	48
total	48	48	48	48	48	48	48	48	384

FIGURE 12.11: IDeC$_{order}$ optimal – Task Instance Count

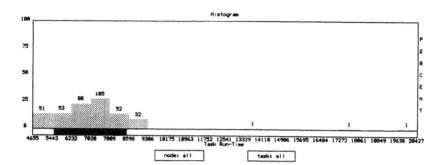

FIGURE 12.12: IDeC_{order} optimal – Task Run Time Histogram (all tasks)

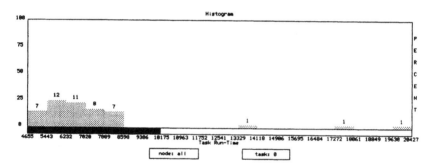

FIGURE 12.13: IDeC_{order} optimal – Task Run Time Histogram (basic step)

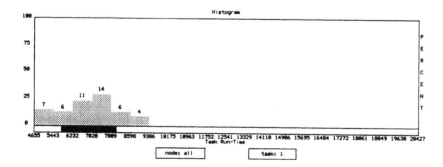

FIGURE 12.14: IDeC_{order} optimal – Task Run Time Histogram (defect correction)

12.1.2   Degenerated Parameters

For the visualization of $IDeC_{order}$ methods with the *degenerated* parameter set,
both the number of defect correction steps required for each interval and the
probability for step size rejections was increased.

The *Utilization Gantt Chart* (Figure 12.15) visualizes two aspects. First,
three processors have to work twice as long as the others. This is due to the
chosen number of required defect correction steps which is not a multiple of
the processor number. Second, the remaining processors (except processor 0)
have long idle periods which are caused by the additional time processor 0
needs if a step size rejection occurs. Both aspects become more apparent in
the *Task Gantt Chart* display (Figure 12.16). Processors 5, 6 and 7 each have
to calculate two defect correction steps for one interval alternately.

The *Space-Time Diagrams* (Figures 12.17, 12.18) reveal the poor performance. Especially obvious are those messages that processor 4 has to send to
processor 5. Since processor 5 has to calculate two defect correction steps for
each interval and processor 4 needs to calculate only one defect correction step
for the same interval, processor 4 sends the messages that contain the solution
matrices much earlier than processor 5 asks for it. The *Space-Time Diagrams*
prove that the time between the sending and receipt of messages corresponding
to these two nodes grows significantly during the course of the program's execution. Compared to the discrepancy between the number of processors and
the number of defect correction steps, the effect of the higher step size rejection
probability may be ignored.

Also the *Utilization Summary* (Figure 12.19) points out the poor load balance. While three processors are utilized for 80 percent of the time, the others
are not even utilized for half of the time. It is true that the utilization for
processor 0 is a little bit higher than that, but the reason for this is just the
increased number of step size rejections.

The *Concurrency Profile* (Figure 12.20) tells the same story. Only for 45
percent of the whole run time are 5 or more processors busy. The highest peak
of this profile (3 processors busy during 25 percent of the run time) stems from
the fact that three of the processors have to calculate two defect correction
steps and are still busy during the last quarter of the execution.

The average run time (Figure 12.21) and the standard deviation run time
(Figure 12.22) of the defect correction step task instances are approximately the
same as with the optimal parameter set (Figures 12.9, 12.10). The increased
step size rejection probability, however, yields additional average run time and
increased standard deviation of the basic step. Also worth a look are the
Task Run Time Histograms (Figures 12.24, 12.25, 12.26), which display the
different run time distributions of the basic step and one defect correction step.

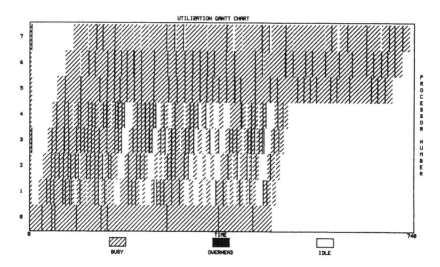

FIGURE 12.15: IDeC_order degenerated – Utilization Gantt Chart

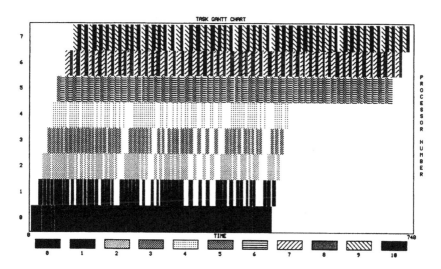

FIGURE 12.16: IDeC_order degenerated – Task Gantt Chart

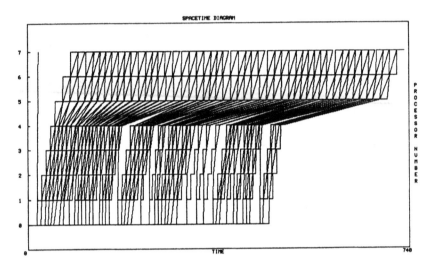

FIGURE 12.17: IDeC$_{order}$ degenerated – Space-Time Diagram

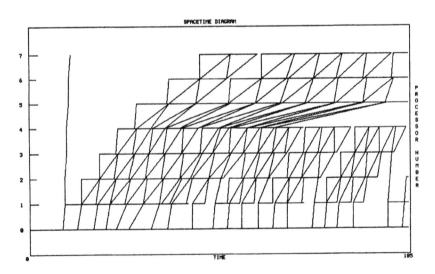

FIGURE 12.18: IDeC$_{order}$ degenerated – Space-Time Diagram (start up)

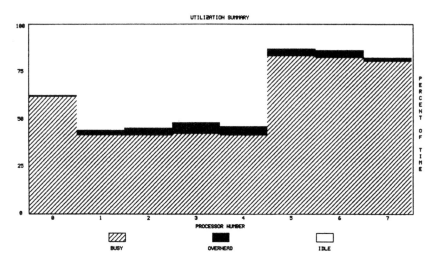

FIGURE 12.19: IDeC_order degenerated – Utilization Summary

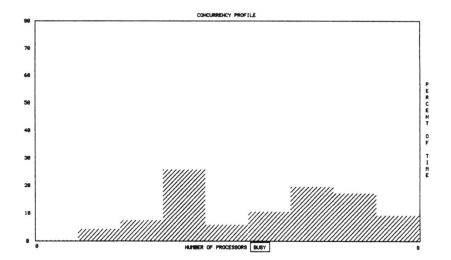

FIGURE 12.20: IDeC_order degenerated – Concurrency Profile

	task0	task1	task2	task3	task4	task5	task6	task7	task8	task9	task10	total
node0	10373											10373
node1		6828										6828
node2			6962									6962
node3				6982								6982
node4					6945							6945
node5						6841	6958					6899
node6								6704	6993			6848
node7										6721	6952	6837
total	10373	6828	6962	6982	6945	6841	6958	6704	6993	6721	6952	7205

FIGURE 12.21: IDeC$_{order}$ degenerated – Task Instance Mean Run Time

	task0	task1	task2	task3	task4	task5	task6	task7	task8	task9	task10	total
node0	4861.2											4861.2
node1		1118.6										1118.6
node2			1157.1									1157.1
node3				1137.1								1137.1
node4					1166.6							1166.6
node5						1287.3	1076.1					1187.9
node6								1290.6	971.14			1151.2
node7										1321.2	1023.5	1187.4
total	4861.2	1118.6	1157.1	1137.1	1166.6	1287.3	1076.1	1290.6	971.14	1321.2	1023.5	2094

FIGURE 12.22: IDeC$_{order}$ degenerated – Task Instance Standard Deviation Run Time

	task0	task1	task2	task3	task4	task5	task6	task7	task8	task9	task10	total
node0	48											48
node1		48										48
node2			48									48
node3				48								48
node4					48							48
node5						48	48					96
node6								48	48			96
node7										48	48	96
total	48	48	48	48	48	48	48	48	48	48	48	528

FIGURE 12.23: IDeC$_{order}$ degenerated – Task Instance Count

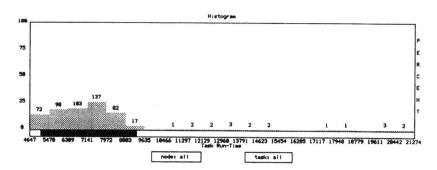

FIGURE 12.24: IDeC_order degenerated – Task Run Time Histogram (all tasks)

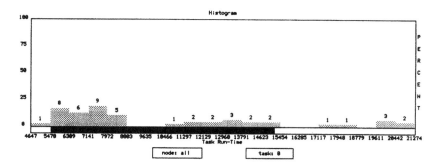

FIGURE 12.25: IDeC_order degenerated – Task Run Time Histogram (basic step)

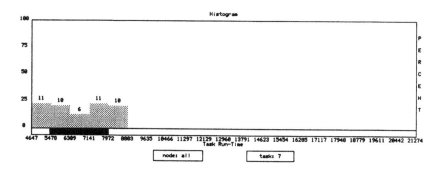

FIGURE 12.26: IDeC_order degenerated – Task Run Time Histogram (defect correction)

12.2 IDeC$_{time}$ Methods

12.2.1 Optimal Parameters

The IDeC$_{time}$ *Utilization Gantt Chart* display (Figure 12.27) shows fewer overhead periods than the corresponding display of IDeC$_{order}$ methods. Since every processor calculates the basic step *and* all defect correction steps of its current interval, only the solution matrices of the basic step need to be passed to the next processor. In that way, communication time can be reduced considerably.

In the *Task Gantt Chart* display (Figure 12.28) basic steps and different defect correction steps may be distinguished from each other by different drawing patterns. The basic idea of IDeC$_{time}$ methods can be seen quite clearly.

The *Space-Time Diagrams* (Figures 12.29, 12.30) prove that the number of messages per interval is the same as on the Sequent Balance and not as high as on the Hypercube performing IDeC$_{order}$ algorithms.

Processor utilization is about 75 percent (*Utilization Summary*, Figure 12.31); the overhead time is negligible.

The *Concurrency Profile* (Figure 12.32) differs from the corresponding display of IDeC$_{order}$ methods (Figure 12.6). For 75 percent of the time at least 6 processors are busy.

Statistics concerning task instance run times are visualized in the *Task Instance Mean Run Time* display (Figure 12.33), *Task Instance Standard Deviation Run Time* display (Figure 12.34), and *Task Instance Count* display (Figure 12.35). In addition, *Task Run Time Histograms* for all tasks (Figure 12.36), for the basic step (Figure 12.37) and for one defect correction step (Figure 12.38) are given.

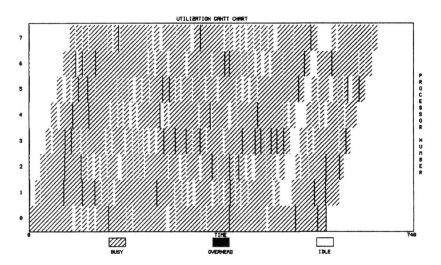

FIGURE 12.27: IDeC$_{\text{time}}$ ideal – Utilization Gantt Chart

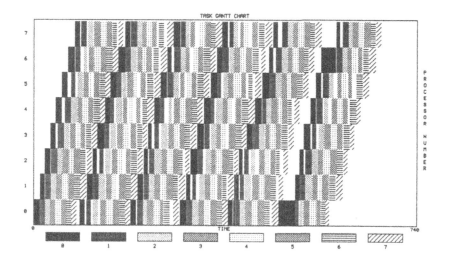

FIGURE 12.28: IDeC$_{\text{time}}$ ideal – Task Gantt Chart

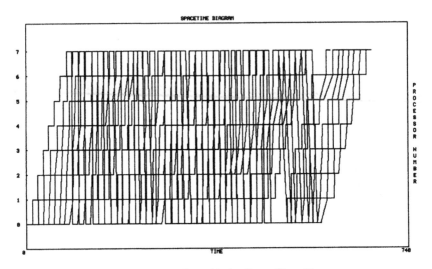

FIGURE 12.29: IDeC$_{time}$ ideal – Space-Time Diagram

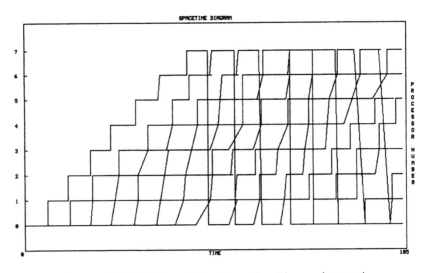

FIGURE 12.30: IDeC$_{time}$ ideal – Space-Time Diagram (start up)

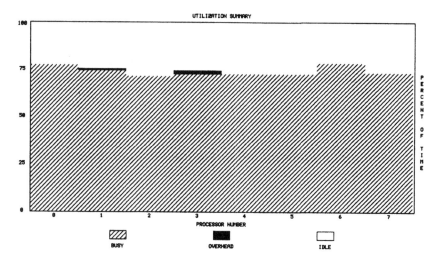

FIGURE 12.31: IDeC$_{time}$ ideal – Utilization Summary

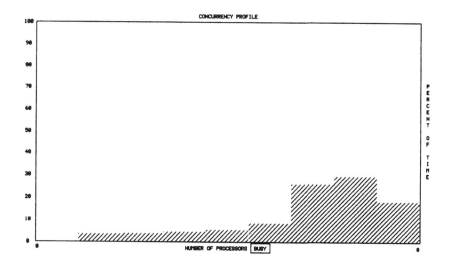

FIGURE 12.32: IDeC$_{time}$ ideal – Concurrency Profile

	task0	task1	task2	task3	task4	task5	task6	task7	total
node0	9082	6809	6337	7144	7325	6970	6461	6414	7068
node1	6065	6357	7383	7198	6184	7323	6830	7614	6869
node2	5868	6859	6735	6692	6430	7398	6223	5574	6472
node3	7124	6716	6634	6908	6357	6512	7282	6197	6716
node4	7109	6115	6616	6586	6149	6545	6968	6960	6631
node5	7336	6810	6131	6321	6095	6145	6937	7077	6607
node6	8674	7734	6172	6172	6978	6458	6992	6895	7009
node7	6616	6856	5745	6339	6354	7196	5927	7471	6563
total	7234	6782	6469	6670	6484	6818	6702	6775	6742

FIGURE 12.33: IDeC$_{\mathrm{time}}$ ideal – Task Instance Mean Run Time

	task0	task1	task2	task3	task4	task5	task6	task7	total
node0	5228.6	1117.2	1117.9	884.83	827.67	1323.1	963.59	1071.3	2254.6
node1	942.4	1145.1	723.68	756.28	878.3	1150.3	1218.8	891.11	1128.2
node2	1424.4	1197.9	1109.3	951.25	742.26	1071.2	1162.1	525.85	1187.7
node3	1297.8	1392.6	839.69	997.53	618.56	936.44	787.61	771.35	1046.9
node4	1047.9	1286.2	1382.8	1030	943.6	737.44	1331	1129.3	1181.3
node5	630.9	1381.3	890.03	952.61	1395	988.34	1077.3	589.09	1124.8
node6	4436.4	703.15	873.3	906.25	968.36	906.99	930.94	962.8	1947.2
node7	958.41	1078	797.42	1185.8	1055.5	640.14	618.6	874.17	1076.7
total	2810.2	1260	1090.4	1030.7	1038	1081.6	1118.9	1081.7	1449.3

FIGURE 12.34: IDeC$_{\mathrm{time}}$ ideal – Task Instance Standard Deviation Run Time

	task0	task1	task2	task3	task4	task5	task6	task7	total
node0	6	6	6	6	6	6	6	6	48
node1	6	6	6	6	6	6	6	6	48
node2	6	6	6	6	6	6	6	6	48
node3	6	6	6	6	6	6	6	6	48
node4	6	6	6	6	6	6	6	6	48
node5	6	6	6	6	6	6	6	6	48
node6	6	6	6	6	6	6	6	6	48
node7	6	6	6	6	6	6	6	6	48
total	48	48	48	48	48	48	48	48	384

FIGURE 12.35: IDeC$_{\mathrm{time}}$ ideal – Task Instance Count

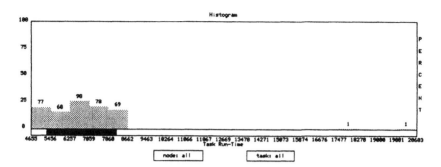

FIGURE 12.36: IDeC$_{\text{time}}$ ideal – Task Run Time Histogram (all tasks)

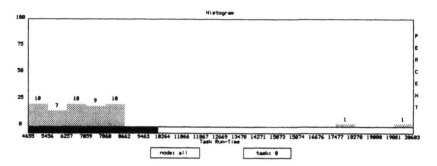

FIGURE 12.37: IDeC$_{\text{time}}$ ideal – Task Run Time Histogram (basic step)

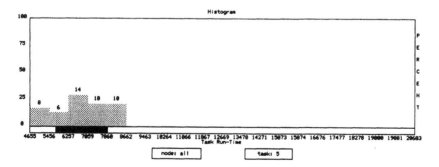

FIGURE 12.38: IDeC$_{\text{time}}$ ideal – Task Run Time Histogram (defect correction)

12.2.2   Degenerated Parameters

For the visualization of IDeC$_{time}$ methods with the *degenerated* parameter set, both the number of the intervals and the probability for step size rejections was increased.

The problem that arises when using a number of intervals that is *not* a multiple of the number of processors is best seen in the *Utilization Gantt Chart* display (Figure 12.39) at the far right. After each processor has completed six intervals there are just four intervals left, which are calculated by the first four processors while the remaining four processors stay idle.

A more serious problem is caused by step size rejections, whose probability is set at 60 percent. Upon examination of the *Task Gantt Chart* display (Figure 12.40) it becomes apparent that the overall performance suffers from this high step size rejection probability which increases the average duration of the basic step.

The *Space-Time Diagrams* (Figures 12.41, 12.42) show the reason for the long idle periods. After the termination of the first interval, processor 0 has to wait for a message (from processor 7) that contains the solution matrix of the basic step. However, this message is delayed by the large number of step size rejections.

The *Utilization Summary* (Figure 12.43) and the *Concurrency Profile* (Figure 12.44) report the same behavior. Processor utilization is only between 50 and 60 percent. 80 percent of the time just 4 to 6 processors are busy, while the state when all processors work concurrently almost never occurs.

Due to the higher step size rejection probability, average task instance run time of the basic step is nearly twice as much as the average task instance run time of one defect correction step (Figure 12.45). For the same reason the standard deviation of the task instance run time of the basic step is much higher than the same statistical value for any defect correction step (Figure 12.46). The *Task Instance Count* display (Figure 12.47) again shows that half of the processors have to calculate one additional interval.

The *Task Run Time Histograms* (Figures 12.48, 12.49, 12.50) visualize the differences in the distribution of run times of the basic step and a defect correction step.

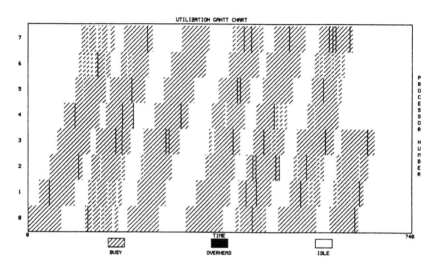

FIGURE 12.39: IDeC_{time} worst case – Utilization Gantt Chart

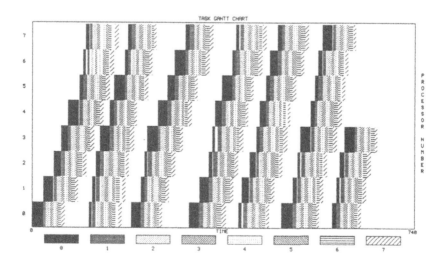

FIGURE 12.40: IDeC_{time} worst case – Task Gantt Chart

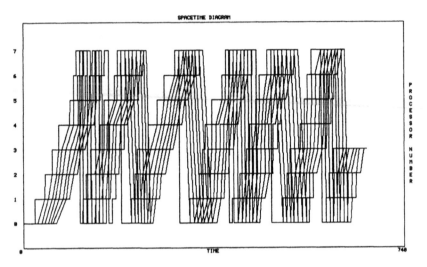

FIGURE 12.41: IDeC$_{time}$ worst case – Space-Time Diagram

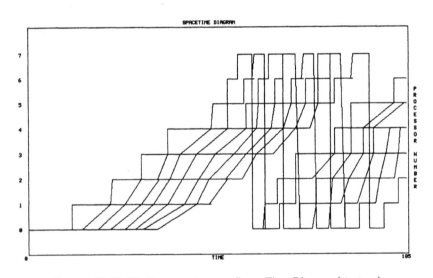

FIGURE 12.42: IDeC$_{time}$ worst case – Space-Time Diagram (start up)

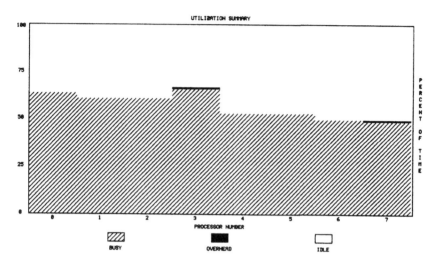

FIGURE 12.43: IDeC$_{time}$ worst case – Utilization Summary

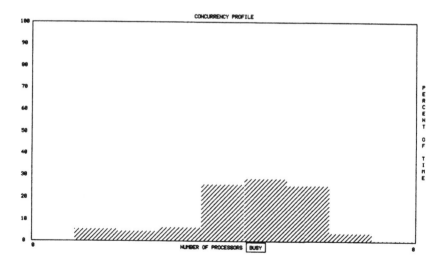

FIGURE 12.44: IDeC$_{time}$ worst case – Concurrency Profile

	task0	task1	task2	task3	task4	task5	task6	task7	total
node0	14268	7022	7116	6480	6295	6561	6501	6408	7581
node1	11344	7371	6607	6105	6662	6698	6902	6532	7278
node2	8668	7491	6355	7129	7314	6466	7004	7138	7196
node3	14776	6398	6865	7298	6375	6324	6744	7342	7765
node4	13252	6260	6191	6911	6071	6627	6370	6440	7265
node5	12421	7060	6570	6848	6494	6680	5768	6490	7291
node6	10271	6344	6173	5699	6134	6688	6940	6857	6888
node7	8956	6611	6083	6550	6993	6652	6246	6813	6863
total	11785	6839	6513	6637	6551	6581	6577	6760	7280

FIGURE 12.45: IDeC_{time} worst case – Task Instance Mean Run Time

	task0	task1	task2	task3	task4	task5	task6	task7	total
node0	6864.8	861.5	882.05	535.09	793.38	1138.9	1119.9	1000.1	3619.7
node1	5546.8	746.85	1418.2	758.8	1310.6	1032.6	948.55	476.38	2683.4
node2	3299.8	1013.2	1133.2	1155.8	1022.5	1120.1	1074.3	689.46	1659
node3	6380.9	1255.6	1184.7	1262.2	993.4	690.21	1340.8	1252.5	3664.1
node4	4971.7	1122.9	1258	1424.2	1287.6	1444.5	1028.4	1158	3106
node5	5961	1001.5	1082.2	1015.6	1176.4	992.01	661.1	1181.2	3041.3
node6	5715	1232.2	1157	743.49	1247.5	1226.2	1216.3	902.06	2638.7
node7	4549.1	914.99	764.13	1129.4	821.29	1338.4	919.04	1240	2056.6
total	5949.2	1123.8	1180.7	1150.4	1166.4	1140.8	1131.2	1066.6	2909.5

FIGURE 12.46: IDeC_{time} worst case – Task Instance Standard Deviation Run Time

	task0	task1	task2	task3	task4	task5	task6	task7	total
node0	7	7	7	7	7	7	7	7	56
node1	7	7	7	7	7	7	7	7	56
node2	7	7	7	7	7	7	7	7	56
node3	7	7	7	7	7	7	7	7	56
node4	6	6	6	6	6	6	6	6	48
node5	6	6	6	6	6	6	6	6	48
node6	6	6	6	6	6	6	6	6	48
node7	6	6	6	6	6	6	6	6	48
total	52	52	52	52	52	52	52	52	416

FIGURE 12.47: IDeC_{time} worst case – Task Instance Count

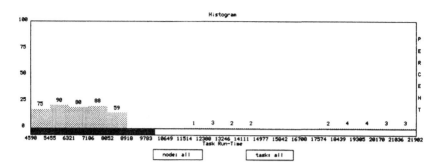

FIGURE 12.48: IDeC$_{\text{time}}$ worst case – Task Run Time Histogram (all tasks)

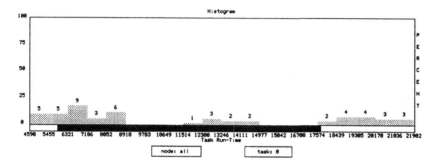

FIGURE 12.49: IDeC$_{\text{time}}$ worst case – Task Run Time Histogram (basic step)

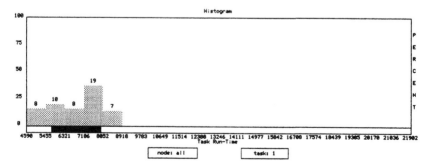

FIGURE 12.50: IDeC$_{\text{time}}$ worst case – Task Run Time Histogram (defect correction)

12.3 IDeC$_{\text{adaptive}}$ Methods

On the Sequent Balance the best results of the visualized IDeC algorithms has been obtained with IDeC$_{\text{adaptive}}$ methods. The basic idea that each processor would ask for a new interval every time it became idle was easy to implement on a shared memory architecture. However, the same concept may not be adapted directly to a message passing system like the Intel Hypercube. Any communication routine needs much more time than on the Sequent Balance; thus the required time that each processor needs to figure out which basic step or which defect correction step of an interval should be calculated next becomes unfeasibly long.

The other Hypercube programs know in advance where to send their solution matrices. This would not be true for IDeC$_{\text{adaptive}}$ algorithms. Consequently one processor would need to become a master which would be responsible for saving the solution matrices of the whole system and for distributing the computational work. A processor (worker) that became idle would have to send a message to that master. The master in turn would return another message containing two solution matrices (one of the basic step and one of the prior defect correction step). After the worker had finished its calculations, it would send the new solution matrix of the defect correction step it had just finished to the master, and ask for another piece of work.

All together this concept would require many more messages than any other method *and* would waste the processor required as the master. Due to the resulting poor processor utilization such an algorithm was not implemented. A different approach is a matter for future research.

Summary

In this part of the book several IDeC algorithms have been visualized to demonstrate the influence of different hardware platforms and algorithm variants.

All of the investigated algorithms achieve good performance as long as the user adjusts the parameters accordingly: both $IDeC_{order}$ and $IDeC_{time}$ algorithms achieve the best speed-ups when the number of processors is equal to the number of working steps (i.e. the number of defect correction steps plus 1 for the basic step). If this is not possible $IDeC_{order}$ algorithms demand that the number of processors divide the number of working steps. Disobeying this rule leads to poor performance. $IDeC_{time}$ algorithms may use any number of processors that is not larger than the number of working steps. This choice influences only the execution time but not the speed-up. Additionally, $IDeC_{time}$ algorithms achieve better performance if the number of intervals to be calculated is a multiple of the number of processors. However, this aspect is not as critical for the performance and the speed-up as the number of processors for $IDeC_{order}$ algorithms.

Overhead times may be ignored on the Sequent Balance, due to the fact that data structures are passed through the shared memory. On the Hypercube, however, messages that contain larger data structures (matrices) have to be passed through the communication system. The resulting higher overhead is especially apparent in $IDeC_{order}$ algorithms since the number of messages to be handled is double that of $IDeC_{time}$ algorithms.

$IDeC_{adaptive}$ algorithms decide dynamically which processor calculates which intervals. On a shared memory computer, where communication time is rather cheap, some time may be spent on communication tasks concerning this decision as long as all processors are utilized to the greatest possible extent. This is the reason why $IDeC_{adaptive}$ algorithms produce the best results on the Sequent Balance. Unfortunately, communication delays are much longer on message passing systems like the Intel Hypercube, as already discussed in the previous chapter. This aspect makes it difficult to implement a straightforward adaptation of the $IDeC_{adaptive}$ algorithm for the Hypercube architecture.

Part III

Visualization of
Parallel Integration Methods

This part of the book describes the authors' experience concerning the visualization of parallel integration programs. Chapter 13 contains an introduction to the parallel integration methods used. The properties of these parallel integration methods were studied by using a special software system capable of simulating arbitrary message-passing parallel machines (Chapter 14). Each simulated parallel program run produces a trace file (Chapter 15) which is subsequently used by PARAGRAPH (Chapter 5, Chapter 16) as an input for post-mortem visualizations. Again, PARAGRAPH was chosen because of its variety and extendibility.

Specific examples of the visualization of different parallel integration methods as well as an evaluation of those methods can be found in Chapter 17.

Chapter 13

Parallel Integration

Parallel integration techniques utilize multi-processor machines to compute numerical approximations Q to definite one- and multi-dimensional integrals. This book only discusses one-dimensional integrals

$$\mathrm{I}f := \int_a^b f(x)\,dx \tag{13.1}$$

for given input data a, b, and f. A numerical approximation Q is supposed to satisfy the error tolerance

$$|Q - \mathrm{I}f| \le \tau, \quad \tau := \max\{\varepsilon_{abs}, \varepsilon_{rel} \cdot |Q|\}, \tag{13.2}$$

where ε_{rel} and ε_{abs} are the requested relative and absolute error bounds, respectively.

The visualized program DEM is a special version of the QUADPACK (Piessens et al. [196]) subroutine QAG parallelized at the subdivision level by using a synchronous multiple task pool method with a central control mechanism (Krommer, Ueberhuber [194]).

The meta-algorithms in Tables 13.1 and 13.2 describe the principal algorithmic structure of DEM. Before sending out load balancing commands, the master determines the worker with the smallest load, wlmin, and the worker with the highest load, wlmax. The load of a worker is defined to be the sum of all local error estimates, i.e. the sum of the error estimates of all locally stored subintervals.

Then, the master broadcasts the identification numbers of these two workers and the amount of work wldif := (wlmax − wlmin)/2 which has to be transferred between them to balance load.

Upon receiving this message, all workers different from those with maximum and minimum load simply continue their work. The worker with the maximum load extracts intervals from its local interval collection until the sum of the error estimates of these intervals is roughly equal to wldif. Then, it sends

master
    receive problem parameters;
    perform initializations;
    apply single quadrature rule to the original interval;
    store overall integral and error estimate;
    if *problem error requirement is not met* **then**
        start workers;
        work_loop:
        do
            do M times
                do L times
                    perform **integration**
                end do
                receive integral and error estimate corrections from workers;
                update overall integral and error estimate;
                if *error requirement is met* **then exit** work_loop
            end do
            send out load balancing commands;
            perform load balancing;
        end do work_loop
    broadcast termination message to all workers;
end master

worker
    perform initializations;
    work_loop:
        do
            do M times
                do L times
                    perform **integration**
                end do
                send integral and error estimate corrections to master;
                if *termination message has been received* **then exit** work_loop
            end do
            receive load balancing commands;
            perform load balancing;
        end do work_loop
end worker

TABLE 13.1: Algorithmic Structure of DEM – Part I

integration
>*fetch subinterval with the largest error estimate*
>>*from the local interval collection;*
>*subdivide this interval;*
>*apply single quadrature rule to new intervals;*
>*insert integral and error estimates in local interval collection;*

end integration

TABLE 13.2: Algorithmic Structure of DEM – Part II

the extracted intervals to the worker with the minimum load and continues its work. The worker with the minimum load receives these intervals, stores them locally, and continues its work.

At the beginning of program execution, DEM obtains the following data:

1. the left integration boundary a;
2. the right integration boundary b;
3. the absolute error requirement ε_{abs};
4. the relative error requirement ε_{rel};
5. the parameter **key**, which determines the specific quadrature rule to be used;[1]
6. the number **nw** of workers;
7. the number L of integration steps between two checks for global convergence;
8. the number M of global convergence steps between two load balancing activities.

---

[1]This parameter may be any integer between 1 and 6 and is associated with Gauss-Kronrod formulas which differ from each other by the number of integration abscissas.

Chapter 14

Simulated Target Machines

The target machine of the visualized parallel integration programs is a *simulator*. While being executed on a sequential computer, the simulator can be tuned to behave like an arbitrary parallel computer.

The simulation system assumes that the underlying multiprocessor system is based on the MIMD principle, i.e. that each processor-memory node has its own instruction stream. Consequently, different processors can execute their tasks independently (asynchronously) of each other. However, in the initial version of the simulation system, an SPMD (Single Program Multiple Data)[1] programming model is used.

For the current simulation system, the *message passing paradigm* for parallel programming was chosen. It is assumed that a system software layer enables processors to communicate directly with each other, i.e. no *explicit* routing has to be carried out to establish communication channels between two nodes that are not directly linked to each other.

The input of the simulator is a Fortran program, which contains special constructs to express concurrency. A preprocessor modifies this input program resulting in another Fortran program that may be compiled with any Fortran 77 compiler. The output program simulates the execution of the input program on parallel hardware.[2]

There exist a great number of parameters, with which communication delays can be quantified. For the visualization of parallel integration programs the simulator was adapted to read three integer variables $i1$, $i2$ and $i3$ from standard input at the program start. These values are used to define the main

---

[1]In contrast to the SIMD model, the SPMD model does not require that all processors be controlled by a single instruction stream. The SPMD model of computation only specifies that all instruction streams must originate from a single program.

[2]The simulation system does *not* attempt to model the performance of sequential program sections on processor-memory nodes. Instead, the simulation system requires the user program to provide the execution times of sequential code sections and to advance processor clocks accordingly whenever sequential code sections are performed.

characteristics of the simulated hardware. $i1$ is the time required to process received messages (i.e. the CPU time spent in a rcv subroutine after a receive waking event). $i2$ is the time required to process messages to be sent (i.e. the CPU time spent in a snd subroutine). $i3$ is the transmission delay (i.e. the time between the end of message processing by the sending processor and the earliest time when the message processing at the receiving processor can begin). For the visualizations found in Chapter 17 all three quantities are assumed to be independent of the message length and of the processors involved.

A detailed discussion of all aspects concerning this simulator can be found in Krommer, Ueberhuber [195].

The advantage of using the simulator software instead of a "real" parallel machine is its high flexibility. The performance of various algorithmic schemes and variants for a representative set of different machine parameters and problem settings may be simulated. Only a few parameters need to be changed to simulate different kinds of current or future parallel computers.

Chapter 15

Trace File

Since the parallel integration programs are executed on a simulator (Chapter 14), the PICL features (Section 9.1) for automatically generating a trace file cannot be used. Thus the creation of a PICL trace file requires either the implementation of a "machine" dependent part written for the simulator, or a modification of the executed sequential programs and/or the simulator in such a way that they *explicitly* create the appropriate trace files.

For the visualization of parallel integration programs, the second method was chosen. This decision also enables the definition of new event types which are needed for passing program information to the newly created *User Gantt Displays* (Section 16). Correspondingly, three new event record types, called DEFINE_USER_GANTT, DEFINE_USER_CLASSES and USER_GANTT, have been introduced.

15.1   DEFINE_USER_GANTT

The DEFINE_USER_GANTT record is needed only once for the definition of a *User Gantt Display*. It has the following format:

52 $time_{high}$ $time_{low}$ type $value_{zero}$ name $items_y$ $label_{node}$

The *name* and the $label_{node}$ parameter are string values.[1] The parameter $value_{zero}$ is a double precision floating point number, all other parameters are integer values. The individual parameters have the meaning described below.

After the record-type, 52, the time of the event has to be specified. Since there is no associated time for this event type, a zero time stamp may be used.[2] The next parameter determines the newly defined USER_GANTT *type*. Each type may be defined only once, and for each subsequent USER_GANTT

---

[1] If a given string value contains blanks, these blanks must be replaced by underscore characters to ensure correct functioning of PARAGRAPH.

[2] These two time values must be given to guarantee conformance with the definition of other PICL trace records.

record (see below) of a particular type a corresponding DEFINE_USER_GANTT record of the same type must exist.   Each time the specified value of the USER_GANTT record is equal to the *zero-value*, the *User Gantt Display* leaves the corresponding space blank.   The given *name* appears in the *user displays* menu of PARAGRAPH.

The last two arguments of the trace record are optional. The number *items_y* (i.e. the number of items displayed on the y axis) is usually equal to the number of processors. In this case this value should be set to -1, the default value. However, if any other item type is to be displayed on the y axis, this parameter defines the respective number of items. By setting the last parameter, the label for the item type may be changed. As a default label the string "Processor Number" is assumed.

Example 1: The trace record

> `52 0 0 1 -99 Error_Estimate`

defines a *User Gantt Display* of type 1. The label "Error Estimate" will appear in the *user displays* menu, and each USER_GANTT record of type 1 with a value of -99 will leave the corresponding display space blank.

Example 2: The trace record

> `52 0 0 2 0 Subintervals`

defines a *User Gantt Display* for showing the number of intervals locally stored at each processor node.

Example 3: The trace record

> `52 0 0 3 -1 Single_Error_Estimate 100 Intervall_Number`

defines a *User Gantt Display* which shows the error estimate of each individual interval. A maximum of one hundred intervals is assumed.

15.2   DEFINE_USER_CLASSES

The DEFINE_USER_CLASSES record is optional.   It may be used to specify the number and type of classes for a single *User Gantt Display*. The specified class partitioning has great influence on the graphical appearance of the *User Gantt Display*. For each class a lower and an upper bound has to be specified. $bound_0$ is the lower bound of the first class, $bound_1$ is the upper bound of the first class and the lower bound of the second class, and so on. PARAGRAPH displays each class in a different color or gray-shade.

53 $time_{high}$ $time_{low}$ type classes scaletype $bound_0$ ... $bound_n$

The parameters for the class *bounds* are double precision floating point numbers; all other parameters are integer values.

This trace record starts with the record type, 53, two time values (usually set to zero), and the associated *User Gantt Display* type as defined in the DEFINE_USER_GANTT record. The next parameter chooses the number of classes. The *scaletype* parameter determines the way the classes are chosen. Possible values are 0, 1, 2 and 9:

Value 0 (automatic): The program heuristically chooses suitable class bounds. If the *value* parameters of the USER_GANTT records for the given type are (approximately) uniformly distributed, classes of equal width are chosen. If this is not true, a logarithmic scale is taken.

Value 1 (linear): The classes are chosen with equal width. The lowest and highest value of the USER_GANTT records determine the lower bound of the first class and the upper bound of the last class, respectively.

Value 2 (logarithmic): Again the minimum value for the first class and the maximum value for the last class are obtained from the USER_GANTT records. However, a logarithmic class partitioning is made.

Value 9 (user-specific): This choice enables the user to specify each single class bound. For this reason num+1 additional parameters are necessary, where num is the specified number of classes. The first additional parameter is the lower bound of the first class; the second parameter is the higher bound of the first class and the lower bound of the second class, etc. The first additional parameter may be a dash, in which case all values that are smaller than the second parameter belong to the first class. Similarly, a dash as the last parameter indicates that all values that are higher than the second last parameter should belong to the last class.

Example 1: The trace record

53 0 0 1 20 2

tells PARAGRAPH to use 20 classes, and that these classes should be partitioned logarithmically.

Example 2: The trace record

53 0 0 2 10 9 0 1 2 3 5 10 20 30 50 100 −

determines a user-specified class partitioning into 10 classes. The individual class bounds are listed as the last eleven parameters. The last parameter guarantees that all "high values" (greater than hundred) belong to the last class.

Example 3: If no associated DEFINE_USER_CLASSES record exists, the number of classes is set to 16 and the scale type is set to 0 (automatic). Thus the specification of the trace record

53 0 0 3 16 0

is equivalent to a missing DEFINE_USER_CLASSES record.

15.3   USER_GANTT

The USER_GANTT record is used to specify the actual values that the user wants to visualize. The format is as follows:

51 time$_{high}$ time$_{low}$ node type value

The *value* parameter is a double precision floating point number; all other parameters are integer values. The trace record starts with the number 51, which characterizes the USER_GANTT record. Two *time* parameters inform PARAGRAPH about the times at which the changes in the visualized programs occurred. The *node* parameter declares the processor to which this record belongs. The *type* of the *User Gantt Display* may be any type previously defined with the DEFINE_USER_GANTT record. The actual *value* to be displayed graphically is expected as the last parameter.

Example 1: The trace record

    51 1 23456 2 1 0.798675

specifies that the display region of processor 2 in the *User Gantt Display* type no. 1 should be colored in a way that depicts the specified value of approximately 0.8. This color is drawn from the specified point of time until the next trace record of the same processor in the same *User Gantt Display* type.

Example 2: The trace record

    51 8 76543 6 2 0

specifies that the display region of processor 6 in the *User Gantt Display* type no. 2 should be left blank until the next occurrence of a trace record for this processor in this *User Gantt Display*.

Chapter 16

Integration-Specific Displays

For the visualization of parallel integration programs, one general purpose display type was implemented. This new display type requires that the trace file contain special trace records, called DEFINE_USER_GANTT, DEFINE_USER_CLASSES and USER_GANTT. The syntax of these records is described in Chapter 15.

User Gantt Displays are capable of showing user-defined metrics graphically. The number of displayable metrics is limited only by the screen size. The visual appearance of this new display type is identical to PARAGRAPH's standard *Utilization Gantt Chart* and *Task Gantt Chart* display. Each of the new displays is arranged so that time is on the horizontal axis and processors[1] are on the vertical axis. The time unit may be chosen with the help of PARAGRAPH's options menu. If the window is not wide enough to display the whole execution history, it scrolls horizontally by the amount specified in the options menu.

The displayed colors or patterns[2] are usually chosen automatically in the following way: after the program start PARAGRAPH checks the given trace file for USER_GANTT trace records. In this way the minimal and maximal values of each user-defined Gantt Chart type can be sorted out. For each type the obtained interval is divided into 16 classes, which are either equally wide or are

---

[1] By default these displays assume that processor numbers are shown on the vertical axis. However, the user is free to choose the name and the number of the items which are displayed on the vertical axis. This is accomplished by using the DEFINE_USER_GANTT trace record (see Section 15.1).

[2] On a color display, a color table is used that ranges from red (high values) through blue (medium) to green (low values). On grayscale displays, these colors are mapped to shades of gray which are hard to distinguish. For this reason PARAGRAPH provides the -m switch which forces all displays to use patterns instead of colors or shades of gray. However, only three different patterns are used repeatedly (the fourth pattern is identical to the first, the fifth pattern is identical to the second, and so on). In the authors' version of PARAGRAPH the new -s switch was introduced. It works similar to the standard -m switch but uses 16 different patterns. This switch may not only be used on monochrome displays but also for the process of collecting screen-hardcopies which will be printing on monochrome output devices. All figures of this book were created with the -s option.

scaled logarithmically depending on the data distribution.

The decision concerning the scale type to be used is accomplished by the program in a heuristic way. If the user does not agree with this choice, he can add a DEFINE_USER_CLASSES trace record to the trace file in order to specify the class partitioning in detail (Section 15.2).

During the visualization process each USER_GANTT trace record maps the given data value to one particular class, whose color/pattern is displayed in the diagram.

One important feature distinguishes *User Gantt Displays* from PARAGRAPH's standard *Gantt Charts*: *User Gantt Displays* may be exposed (redisplayed[3]) and resized *without* losing their displayed information. Standard PARAGRAPH *Gantt Charts* are erased when they are exposed or resized, even if only a part of them needs redrawing.

16.1 Display Categories

The display types of PARAGRAPH which were used in this part of the book can be categorized as follows:

Static Displays (*Utilization Summary* and *Statistical Summary Display*) show certain program states related to the whole program execution.[4]

Gantt Charts (*Utilization Gantt Chart*, *Space-Time Diagram* and *User Gantt Display*) have time on the horizontal axis and the processor number on the vertical axis. The scale of the time axis may be changed by the user even during the visualization. If additional space is required, these displays may scroll horizontally. The scale of the figures in Chapter 17 is chosen in such a way that they either show the entire program run or a certain time interval of special interest.

---

[3]A window may be (partially) obscured by one or more other windows. After it is raised, it becomes fully visible. The information that was hidden below other window(s) has to be redisplayed, which is only possible if the displayed data has been copied to some program buffer.

[4]Both displays can be animated during the program run. However, this dynamic feature is not easy to illustrate on paper and is therefore not dealt with in this book.

Chapter 17

Visualization of Parallel Integration Algorithms

This chapter is divided into several parts. Starting with the discussion of the basic ideas of the parallel integration algorithm, the algorithm is slightly improved in different aspects. In each part of this chapter a specific integration algorithm with a specific parameter set is examined with the aid of different kinds of visualization and compared with previously discussed algorithms and parameter sets.

For the visualizations made in this chapter the parameters studied in Chapter 13 were chosen in the following way: the integration boundaries were set to 0 (left boundary) and 1 (right boundary). The absolute error requirement ε_{abs} was set to 0; the relative error requirement ε_{rel} was set to 10^{-5}. The **key** parameter was set to 1. The number of workers chosen was four, except in Section 17.2 (only one worker). The number L of integration steps between two checks for global convergence was adapted for each algorithm in a way that yielded optimal performance. The number M of global convergence checks between two load balancing activities was set to 1. This means that after each global convergence check a load balancing step is performed. The time needed for the "integration" subroutine was set to exactly 100 (i.e. the corresponding variation was set to 0).

The function f which is integrated by DEM is defined as follows:

$$f(x) = \cos(c\,x^8).$$

The complexity of the integrated function (and thus the run time) is determined by the parameter c. Changes in this parameter result in a more or less oscillating function.

17.1   Basic Algorithm

This section discusses various aspects of the basic parallel integration algorithm. The *Utilization Gantt Chart* (Figure 17.1) and the *Space-Time Diagram* (Figure 17.2) are snapshots of the start-up phase. The diagrams in Figures 17.3 and 17.4 show the same display types at a later time instance. All these figures clearly show two phases that change alternately. First, there is a *calculation phase*, which is labelled **busy** and drawn with slanted lines. During this period a fixed number of intervals is integrated. The second phase, the *communication phase*, is labelled **overhead** and drawn solid. The main reason for this phase is the coordination and processing of a load balancing step.

Apart from these two phases, there are *idle times* (shown as blank regions) which may have two causes. First, not all processors can be utilized from the very beginning. It takes some time to distribute work among processors. A possible solution to this problem would require the time between two load balancing steps to be variable. This feature, however, is still a matter for future research.

The second reason for idle times arises when a processor waits for a responding message that gives it details about the load balancing process. Section 17.3 describes a possible solution to this problem.

The flow of messages is visualized in the *Space-Time Diagrams*. At the very beginning, processor 0 (i.e. the master) broadcasts the *initialization message* to each worker (i.e. all other nodes). The workers receive this message after a short delay and start working on the integration of their locally stored intervals. In the beginning the workers are not assigned any intervals. Therefore, each worker returns a message immediately. Note that a worker is not able to respond directly after the message from the master has arrived. Some time is required to process the incoming message. For a similar reason PARAGRAPH's *Space-Time Diagrams* continue the horizontal line after a worker's message is sent because this time interval is required to handle an outgoing message.

The master integrates a fixed number of intervals before receiving the messages from the workers. Again, note the time between the receipt of the four incoming messages which is required for internal processing.

Since the overall error requirement is usually not met after the first *calculation phase*, the master broadcasts the *continue message*. This messages is directly followed by the *load balancing message*, which contains the indices of those two processors that should transfer intervals. These two processors have to send or receive an additional message which contains the transferred intervals before they start the *calculation phase*. Processors not involved in the load balancing process may start with the *calculation phase* immediately after receiving the *continue message*.

The *Space-Time Diagram* (Figure 17.2) reports that during the first load balancing phase (LBP) processor 0 transfers intervals to processor 1. The second LBP moves intervals from processor 0 to processor 2, the third one from processor 1 to processor 3, and the fourth LBP from processor 3 to processor 4, which finishes the start-up phase (i.e. the time until all processors are (at least) partly utilized). The second *Space-Time Diagram* (Figure 17.4) provides information about later load balancing steps. Intervals are transferred between processors 3 and 4, processors 1 and 3, processors 4 and 2, processors 3 and 0, and processors 1 and 2, respectively.

The *Utilization Summary* (Figure 17.5) reveals that the master (processor 0) has many messages to process and therefore yields huge overhead times (33 percent, colored black). The overhead times of the workers are by no means negligible, but only half (17 percent) of the master's overhead time. In this display the workers' columns look like stairs leading downward. This is caused by the delayed start of successive workers. The overall performance is rather poor, total utilization is below 50 percent. However, this first implementation of the integration algorithm is supposed to serve as a basis for different types of improvements that will be discussed in the remainder of this chapter.

The display of the *Relative Error Estimate* (Figure 17.6) requires some explanations: the displayed patterns correspond to relative error estimates of the currently integrated intervals. At any time all intervals (of all processors) can be divided into intervals that are currently being integrated and intervals that are waiting to be processed. From those latter intervals there is one interval with the highest (absolute) error estimate $error_{max}$. Each time the integration of an interval starts, this display graphically depicts the value of the relative error estimate of that interval, which is defined to be its absolute error estimate divided by $error_{max}$.

Thus the value for relative error estimate is between 0 and 1. A value close to 1 signals that an interval with a high error estimate has been chosen. When subdivided and recalculated these intervals reduce the total error estimate by the highest possible amount. Therefore, such values displayed by dark patterns indicate a good load balance. Perfect conditions are visualized by (nearly) black patterns for all processors.

Figure 17.6 displays five *calculation phases*. During the third and the fifth phase, load is balanced properly. During the other phases few processors are calculating "difficult" intervals (i.e. those with a high error estimate), while the remaining processors are integrating "easy" intervals whose error estimate cannot be improved significantly. This is, for instance, true during the first phase, where only processor 4 calculates "difficult" intervals.

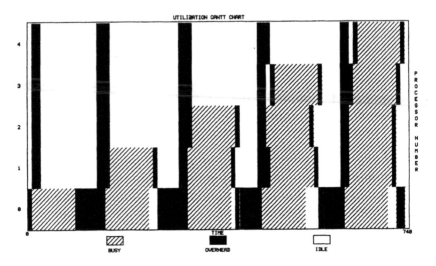

FIGURE 17.1: Basic Algorithm – Utilization Gantt Chart (start up)

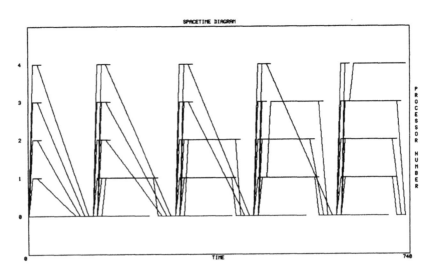

FIGURE 17.2: Basic Algorithm – Space-Time Diagram (start up)

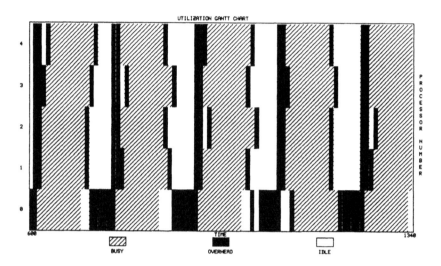

FIGURE 17.3: Basic Algorithm – Utilization Gantt Chart

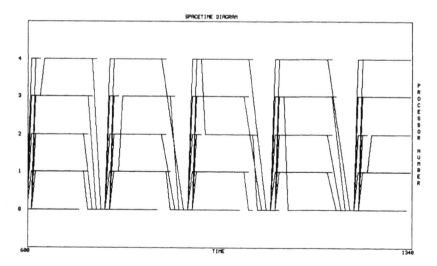

FIGURE 17.4: Basic Algorithm – Space-Time Diagram

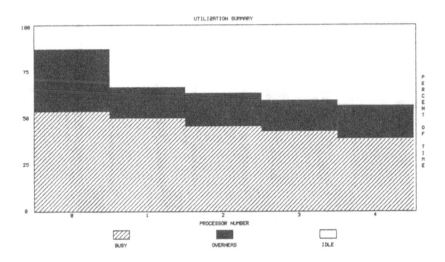

FIGURE 17.5: Basic Algorithm – Utilization Summary

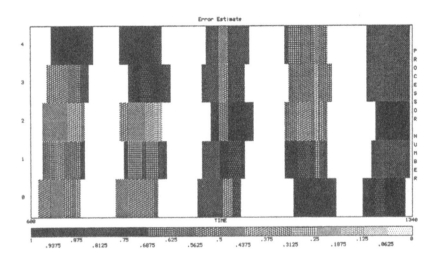

FIGURE 17.6: Basic Algorithm – Relative Error Estimate

17.2 Message Combining

For the experiments in this section, the number of workers was reduced to 1 in order to identify the individual messages more easily. In Section 17.1 it is demonstrated that the master almost always[1] has to broadcast *two* messages in direct succession: the first message contains just a *continue* signal; the second holds information about the load balancing procedure. The obvious idea of combining these two messages to save communication (overhead) time is realized in this section.

Figure 17.7 shows the *Space-Time Diagram* with the conventional algorithm, Figure 17.8 the same diagram with *message combining*. Each *communication phase* requires one message less broadcast. Since the communication time does *not* depend on the message length, the expected speed-up should be notable.

The Utilization Summaries (Figures 17.9, 17.10) demonstrate that the overhead time was reduced from 18 to 15 percent, both for processors 0 and 1.

---

[1] At the very beginning the master broadcasts only the *initialization message*. Similarly, at the very end the master broadcasts only the *termination message*. In both cases no *load balancing message* follows. In all other cases the *load balancing message* immediately follows the *continue message*.

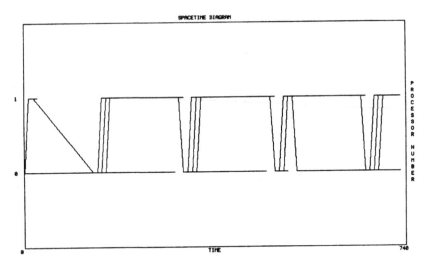

FIGURE 17.7: *Without* Message Combining – Space-Time Diagram

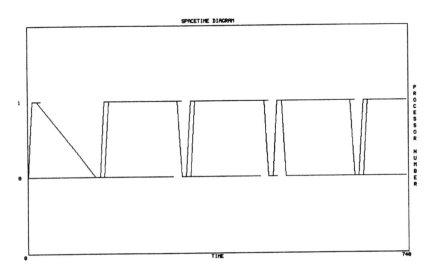

FIGURE 17.8: With Message Combining – Space-Time Diagram

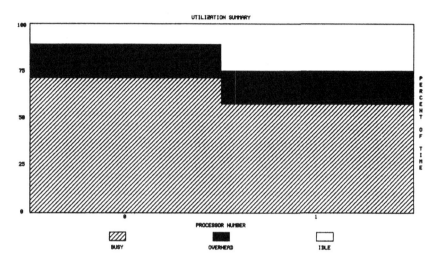

FIGURE 17.9: *Without* Message Combining – Utilization Summary

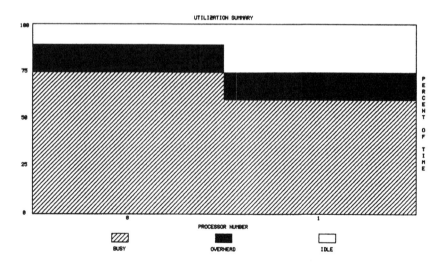

FIGURE 17.10: With Message Combining – Utilization Summary

17.3   Reducing Idle Times

In the *Utilization Gantt Chart* (Figure 17.11) long idle periods (white color) are apparent. The *Space-Time Diagram* (Figure 17.12) reveals that each worker first sends its local integral and error estimate changes to the master and thereafter has to wait for a responding message. This message tells the worker whether or not it should continue working and how it should participate in load balancing actions.

For the improvement made in this section, each worker needs to know approximately the time it takes to get a response from the master after sending it a message. During this estimated time interval, each processor may continue to integrate its locally stored intervals.

The more accurate the assumptions a worker can make, the higher the performance improvements will be. If the actual response time is shorter than the estimated one, the load balancing commands might be received with a delay which may cause workers to integrate "easy" instead of "difficult" intervals. These "difficult" intervals may, accidentally, be clustered on a few processors and have to wait to be distributed among all processors. However, in the experiments, the maximum run time variation (see Chapter 13) was set to zero, which makes it rather easy to predict these response times.

The *Utilization Gantt Chart* (Figure 17.13) and the *Space-Time Diagram* (Figure 17.14) show visualizations of the resulting algorithm. The aforementioned method to avoid idle times was not applied in the master program. The method requires an estimate of the prospective idle times. There is always a danger of overestimating these idle times, which results in a delayed message reception. The load balancing decisions which are based on these messages are also delayed. Since all workers strongly depend on the master, any additional delay causes a performance decline.

The idle times of workers were reduced dramatically. This is a sign that the workers' estimate of the response time was *not* too small. However, this time estimates were also not too large. If this had been the case, broadcasting messages from the master would have been received by the workers with a significant delay, which in turn would have caused the corresponding lines in the *Space-Time Diagram* (Figure 17.14) to lean more to the right than they actually do.

The *Utilization Summaries* for the original algorithm and for the algorithm with reduced idle times are shown in Figure 17.15 and Figure 17.16, respectively. As expected, the column for the master did not change. The changes for the workers, on the other hand, cannot be overlooked. While the amount of the overhead times remained the same, the new algorithm reduced the idle times significantly. The idle times of processor 1 dropped from 25 to 8 percent, those of processor 4 from 39 to 29 percent.

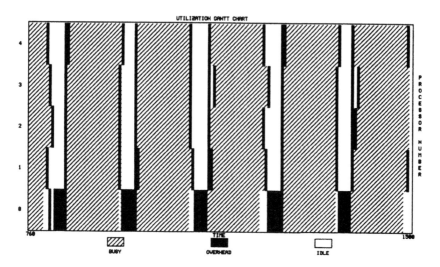

FIGURE 17.11: Original Algorithm – Utilization Gantt Chart

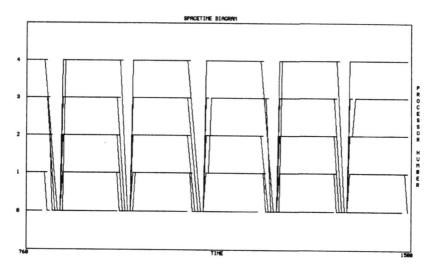

FIGURE 17.12: Original Algorithm – Space-Time Diagram

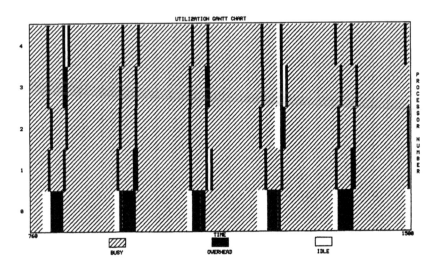

FIGURE 17.13: Reduced Idle Times – Utilization Gantt Chart

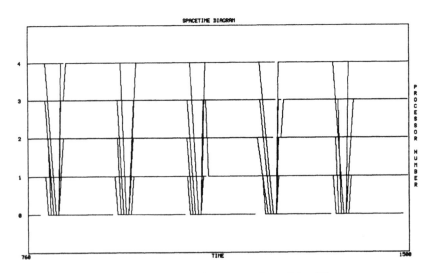

FIGURE 17.14: Reduced Idle Times – Space-Time Diagram

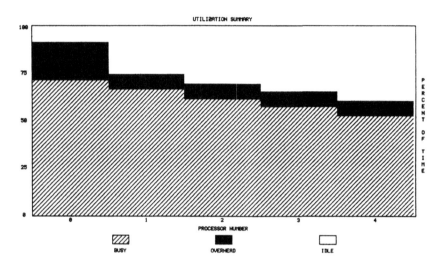

FIGURE 17.15: Original Algorithm – Utilization Summary

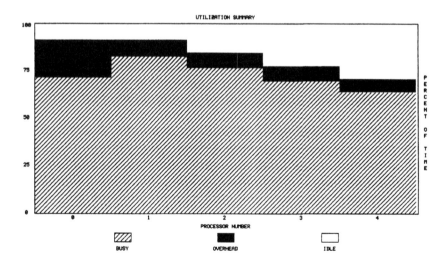

FIGURE 17.16: Reduced Idle Times – Utilization Summary

17.4   Choice of Granularity

An aspect which has not been discussed yet is the proper choice of the algorithmic granularity, i.e. the time between two load balancing steps. This time interval may be controlled with the help of the parameter L (see Chapter 13).

For the first trace file visualized in this section, the parameter L was set to 30. This means that after thirty integration steps a load balancing step is performed. The *Utilization Gantt Charts* (Figures 17.17, 17.18) indicate good performance, which is confirmed by the *Utilization Summary* (Figure 17.19) and the *Statistical Summary Display* (Figure 17.20). The master is utilized up to 79 percent and the sum of its overhead times is only 14 percent of the whole run time. Workers are utilized between 80 and 89 percent and have overhead times of approximately 6 percent each.

In order to reduce the algorithmic granularity, the parameter L was chosen to be 3. The resulting visualizations are shown in Figure 17.21 (*Utilization Gantt Chart*) and Figure 17.22 (*Space-Time Diagram*). The increased number of *communication phases* caused higher communication traffic (i.e. more messages), which in turn resulted in increased overhead times. The *Utilization Summary* (Figure 17.23) and the *Statistical Summary Display* (Figure 17.24) warn the user about the high overhead rates of 48 percent for the master and 21 percent for each worker. Accordingly, busy times are only 28 percent for the master and 71 percent for each worker.

Finally, visualizations of a program with increased load balancing granularity ($L = 300$) can be seen in Figure 17.25 (*Utilization Gantt Chart*) and Figure 17.26 (*Space-Time Diagram*)[2]. Only six load balancing steps took place during the whole program run. This is the reason why only after half of the run time *all* processors are able to integrate intervals. The *Utilization Summary* (Figure 17.27) and the *Statistical Summary Display* (Figure 17.28) show negligible overhead times, caused by the low number of load balancing steps. However, the workers are by far *not* utilized optimally (processor 4 only to an extent of 42 percent).

Another problem that occurs when using such a high load balancing granularity becomes apparent in the *Error Estimate Display* (Figure 17.29). Since the time between two load balancing steps is rather long, the load balance becomes extremely bad. This is displayed by light patterns which indicate small relative error estimates of each processor's currently processed interval.

---

[2]Note that these two displays are scaled in such a way that they display the whole execution history instead of only 10 percent of it, as do the previous displays of this section.

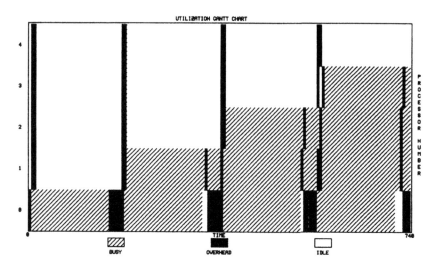

FIGURE 17.17: Medium Granularity ($L = 30$) – Utilization Gantt Chart (start up)

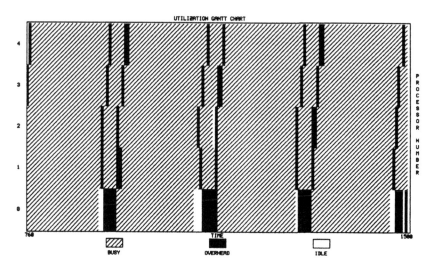

FIGURE 17.18: Medium Granularity ($L = 30$) – Utilization Gantt Chart

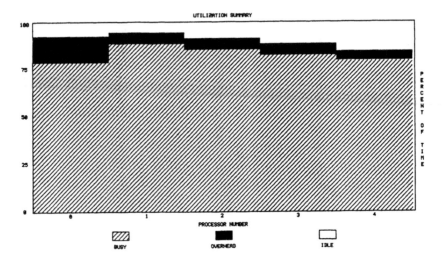

FIGURE 17.19: Medium Granularity ($L = 30$) – Utilization Summary

	Aggregate	node 0	node 1	node 2	node 3	node 4
Percent Processor Busy	83	79	89	86	83	80
Percent Processor Ovhd	7	14	6	6	6	5
Percent Processor Idle	10	7	5	8	11	15
Number Msgs Sent	282	139	39	36	34	34
Total Bytes Sent	29072	9488	3888	11856	2128	1712
Number Msgs Rcvd	281	128	38	42	37	36
Total Bytes Rcvd	29072	4096	11164	5724	4348	3740
Max Queue Size (count)	36	4	2	2	2	2
Max Queue Size (bytes)	8128	864	8056	1504	1024	1048
Max Msg Sent (bytes)	8032	1504	864	8032	800	480
Max Msg Rcvd (bytes)	8032	864	864	1504	1024	1024

FIGURE 17.20: Medium Granularity ($L = 30$) – Statistical Summary

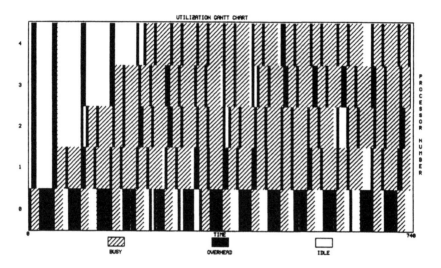

FIGURE 17.21: Small Granularity ($L = 3$) – Utilization Gantt Chart

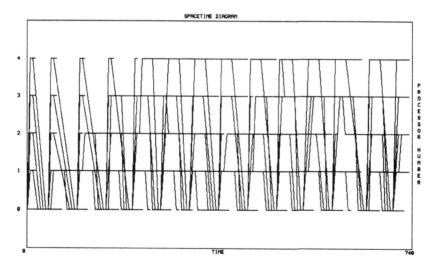

FIGURE 17.22: Small Granularity ($L = 3$) – Space-Time Diagram

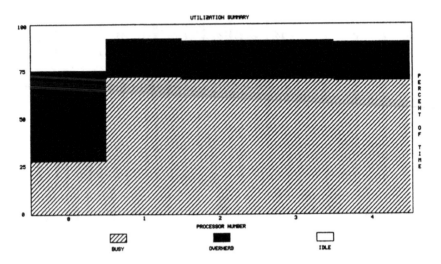

FIGURE 17.23: Small Granularity ($L = 3$) – Utilization Summary

	Aggregate	node 0	node 1	node 2	node 3	node 4
Percent Processor Busy	62	28	72	71	71	70
Percent Processor Ovhd	26	48	21	21	21	21
Percent Processor Idle	12	24	7	8	8	9
Number Msgs Sent	1344	646	171	177	175	175
Total Bytes Sent	49456	18416	5264	10864	8112	6800
Number Msgs Rcvd	1344	613	190	182	183	176
Total Bytes Rcvd	49456	12128	9388	9772	11308	6860
Max Queue Size (count)	154	4	1	2	2	2
Max Queue Size (bytes)	4272	256	352	1856	3456	256
Max Msg Sent (bytes)	3456	192	224	3456	1856	320
Max Msg Rcvd (bytes)	3456	256	352	1856	3456	256

FIGURE 17.24: Small Granularity ($L = 3$) – Statistical Summary

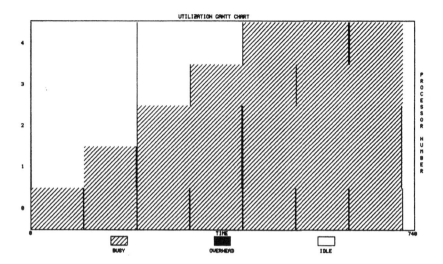

FIGURE 17.25: Large Granularity ($L = 300$) – Utilization Gantt Chart

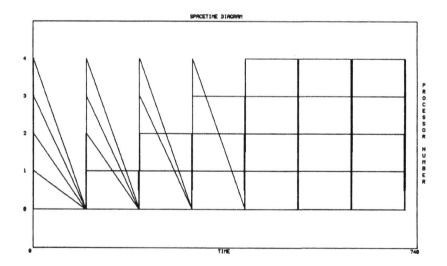

FIGURE 17.26: Large Granularity ($L = 300$) – Space-Time Diagram

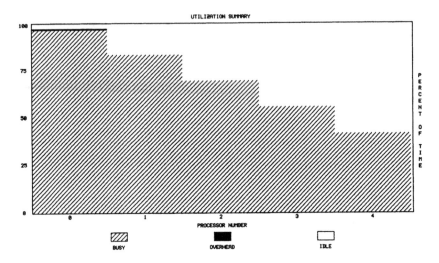

FIGURE 17.27: Large Granularity ($L = 300$) – Utilization Summary

	Aggregate	node 0	node 1	node 2	node 3	node 4
Percent Processor Busy	70	97	84	70	56	42
Percent Processor Ovhd	0	1	0	0	0	0
Percent Processor Idle	30	2	16	30	44	58
Number Msgs Sent	66	37	8	7	7	7
Total Bytes Sent	19248	15056	3856	112	112	112
Number Msgs Rcvd	66	28	9	9	10	10
Total Bytes Rcvd	19248	448	3068	4764	5916	5052
Max Queue Size (count)	12	4	1	1	2	2
Max Queue Size (bytes)	4696	64	2880	4576	3744	3520
Max Msg Sent (bytes)	4576	4576	3744	16	16	16
Max Msg Rcvd (bytes)	4576	16	2880	4576	3744	3520

FIGURE 17.28: Large Granularity ($L = 300$) – Statistical Summary

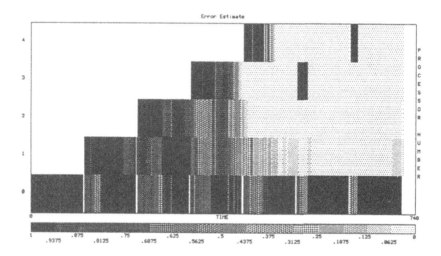

FIGURE 17.29: Large Granularity ($L = 300$) – Error Estimate

Summary

As demonstrated in Chapter 17, the information provided by various visualization techniques significantly facilitates the analysis and improvement of parallel program performance. The basic concept of the investigated algorithm was demonstrated in Section 17.1. Sections 17.2 and 17.3 presented improvements which reduced the number of messages (and thus the overhead times) and others that minimized idle times. Such idle times usually occur whenever a worker waits for a response from the master. Finally, Section 17.4 points out algorithmic granularity given by the time interval between two load balancing phases. However, there still remain many aspects which – when properly solved – may yield notable performance improvements.

For the given algorithms the master is definitely the bottleneck. This prohibits the use of much more than ten processors. If more processors are used, the master is only able to execute a few integration steps and spends most of the time sending and receiving events from all the workers. If *many* more than ten processors are used the master is even not able to respond to incoming messages immediately so that the execution time increases with an increase of processors.

Even with fewer than ten processors each worker has to wait for quite a long time after sending a message to the master until it gets a response. Of course, it has been demonstrated that these originally unused times can be utilized to continue work on locally stored intervals. However, a single worker cannot know if its locally stored intervalls should be integrated next (because they are the ones with the globally largest error estimate), or if intervals from other workers should be transferred first.

Dynamic load balancing should ensure that all processors have the same amount of work. The simple load balancing algorithm realized in the investigated algorithms, i.e. transferring work from the processor with the most work to the one with the least work, cannot always fulfill this demand. But even this simple algorithm requires a decision on *when* these load balancing steps should occur. The visualized algorithms use a fixed number of integration steps between two load balancing steps. This a priori decision rule is easy to implement but may not be adequate in every case. The procedure of determining an optimal value for the number of integration steps between two load balancing steps may be tedious, since it depends on the algorithm as well as on the input data.

Changing the algorithm to a *dynamic* decision when the next load balancing step should take place would solve the problem. Depending on the current load differences, the time span to the next load balancing step should be determined. If there are great load differences (for instance, in the beginning where one processor has the whole interval while the others are idle), the next load balancing step should be performed after only a few integration steps. In contrast, the more equally utilized the processors are, the longer the time interval to the next load balancing step may be.

Bibliography

Parallel Processing

[1] M. A. Arbib, J. A. Robinson (Eds.), *Natural and Artificial Parallel Computation*, MIT Press, Cambridge MA, 1990.

This book is the outcome of a symposium held at Syracuse University in 1989. It contains eleven contributions written by leaders in the fields of neuroscience, artificial intelligence, and cognitive science. The articles cover the phenomena of parallelism in both natural and artificial systems, from the neural architecture of the human brain to the electronic architecture of parallel computers.

[2] J. J. Dongarra, I. S. Duff, D. C. Sorensen, H. A. van der Vorst, *Solving Linear Systems on Vector and Shared Memory Computers*, SIAM Press, Philadelphia, 1991.

The purpose of the book is to unify and document many of the techniques and much of the current understanding about solving systems of linear equations on vector and shared-memory parallel computers. The book is meant to provide a fast entrance to the world of vector and parallel processing for these linear algebra applications. The book is divided into four sections:

1. introduction to terms and concepts, including an overview of the state of the art for high-performance computers and a discussion of performance evaluation;

2. direct solution of dense matrix problems;

3. direct solution of sparse matrix problems; and

4. iterative solution of sparse matrix problems.

The authors have purposely avoided highly detailed descriptions of popular machines and have tried instead to focus as much as possible on concepts; nevertheless, to make the description more concrete, they point to specific computers.

[3] G. C. Fox, M. A. Johnson, G. A. Lyzenga, S. W. Otto, J. K. Salmon, D. W. Walker, *Solving Problems on Concurrent Processors*, Vol. I, Prentice-Hall, Englewood Cliffs, 1988.

The authors derive fundamental computer science insights by analyzing and generalizing the results of explicit concurrent implementations of mainly scientific algorithms. They call their problem-driven approach *computational science*. The book concentrates on practically motivated model problems which serve to illustrate generic algorithmic and decomposition techniques. Only reasonably "regular" problems, for which straightforward techniques can yield good concurrent performance, are covered .

[4] G. A. Geist, M. T. Heath, B. W. Peyton, P. H. Worley, PICL: *A Portable Instrumented Communication Library, C Reference Manual*, Technical Report ORNL/TM–11130, Oak Ridge National Laboratory, 1990.

PICL is a subroutine library that can be used to develop parallel programs that are portable across several distributed memory multiprocessors. PICL provides a portable syntax for key communication primitives and related system calls. It also provides portable routines to perform certain widely-used, high-level communication operations, such as global broadcast and global summation. Finally, the library provides an execution tracing facility that can be used to monitor performance or to aid in debugging. This report is the PICL reference manual for C programmers. It contains full description of all PICL routines as well as explanations on how to use the routines to write parallel programs. A short user's guide is available as a separate report (Geist et al. [5]).

[5] G. A. Geist, M. T. Heath, B. W. Peyton, P. H. Worley, *A user's guide to* PICL: *A Portable Instrumented Communication Library*, Technical Report ORNL/TM–11616, Oak Ridge National Laboratory, 1990.

This user's guide contains an overview of PICL and how it is used; examples in C and Fortran are included. See Geist et al. [4] for a complete reference.

[6] K. B. Kenny, K.-J. Lin, *Measuring and Analyzing Real-Time Performance*, IEEE Software 8–5 (1991), pp. 41–49.

FLEX is a tool developed to measure and analyze the real-time performance. FLEX uses an empirical approach that first measures the actual timing behavior and then uses the measurement results to determine the parameters of a user-supplied timing model. The graphical presentation of the expected behavior gives the user a better understanding about the model.

[7] A. R. Krommer, C. W. Ueberhuber, *Architecture Adaptive Algorithms*, Parallel Computing 19 (1993), pp. 409–435.

The *architecture adaptive algorithm* (AAA) methodology introduced in this paper is an attempt to provide concepts and tools for writing *portable* parallel programs that run *efficiently* on a broad range of target machines. The development of architecture adaptive algorithms and their implementation requires a substantial investment of manpower. Hence, the AAA approach will be most effective in cases where extensive reuse of software is an important characteristic, like, for instance, libraries of scientific parallel programs or large scientific problem solving packages.

First, the concept and the scope of the AAA methodology is outlined. Then a framework for discussing the efficiency of parallel programs in a systematic way is presented, followed by current approaches to achieve efficient parallel software. Finally different aspects of the AAA methodology are discussed and an outlook on future research is given.

[8] M. Kumar, *Measuring Parallelism in Computation-Intensive Scientific/Engineering Applications*, IEEE Transactions on Computers 37 (1988), pp. 1088–1098.

A software tool named COMET was developed for *measuring the total parallelism* in Fortran programs. Total parallelism means that machine-dependent overhead and limitations are filtered out. To measure the parallelism the program P will exhibit while executing on the ideal parallel machine (has unlimited processors and memory and does not incur any overhead), the program P is extended to produce another program P'. The program P' thus obtained is compiled and executed, using one or more input data sets intended to be used with the original program P. The output obtained on executing P' consists of the results expected from the execution of P and a histogram of the parallelism observed in the program. One important conclusion of the article is that parallelism is fairly uneven in most applications.

[9] L. Lamport, *Time, Clocks, and the Ordering of Events in a Distributed System*, Communications of the ACM 21 (1978), pp. 558–565.

The concept of one event happening before another in a distributed system is examined, and is shown to define a partial ordering of the events. A distributed algorithm is given for synchronizing a system of logical clocks which can be used to totally order the events. The use of the total ordering is illustrated with a method for solving synchronization problems. The algorithm is then specialized for synchronizing physical clocks, and a bound is derived on how far out of synchrony the clocks can become.

[10] T. J. LeBlanc, *Parallel Program Debugging*, in "Digest of Papers (COMPCON)" (H. S. AlKhalib, K. Miura, Eds.), IEEE Press, San Francisco, 1989, pp. 65–66.

Problems concerned with *parallel program debugging* are shown. Parallel programs often exhibit nonrepeatable behavior. Interactive debugging can distort the execution and requires the collection and management of an enormous amount of data.

[11] T. J. LeBlanc, J. M. Mellor-Crummey, *Debugging Parallel Programs with* INSTANT REPLAY, IEEE Transactions on Computers 36 (1987), pp. 471–482.

INSTANT REPLAY is a debugger for parallel programs. During program execution the relative order of significant events is saved. The technique guarantees reproducible program behavior during the debugging cycle by using the same input from the external environment and by imposing the same relative order on events during replay that occurred during the original execution. Because of special properties, INSTANT REPLAY is especially useful for debugging parallel programs on tightly coupled multiprocessors, where interprocess communication is cheaper, and therefore more frequent than in loosely coupled systems.

[12] C. E. McDowell, *A practical algorithm for static analysis of parallel programs*, Journal of Parallel and Distributed Computing 6 (1989), pp. 515–536.

One approach to analyzing the behavior of a concurrent program requires determining the *reachable program states*. A program state consists of a set of task states, the values of shared variables used for synchronization, and local variables that derive the values directly from synchronization operations. However, the number of reachable states rises exponentially with the number of tasks and becomes intractable for many concurrent programs. A variation of this approach merges a set of related states into a single *virtual state*. Using this approach, the analysis of concurrent programs becomes feasible as the number of virtual states is often orders of magnitude less than the number of reachable states. The article presents a method for determining the virtual states that describe the *reachable program states*, and the reduction in the number of states is analyzed. The algorithms given have been implemented in a static program analyzer for multitasking Fortran, and the results obtained are discussed.

[13] C. E. McDowell, D. P. Helmbold, *Debugging Concurrent Programs*, ACM Computing Surveys 21 (1989), pp. 593–622.

A survey of *debugging techniques* is presented and problems associated with debugging parallel programs are described. The authors distinguishes four groups of debugging techniques. The *traditional* debugging techniques, also called "breakpoint" debuggers, consist of a collection of sequential debuggers, one per parallel process. The *event based* debuggers view the execution of a parallel program as a sequence of events. *Static analysis* techniques are based on dataflow analysis of parallel programs. And finally techniques for *displaying the flow of control* and distributed data of parallel programs. Further presented are four basic techniques for displaying debugging information:

1. the textual representation of data;
2. the time-process diagrams;
3. the animation of the program execution; and
4. the multiple windows displays.

[14] C. M. Pancake, *Software Support for Parallel Computing: Where Are We Headed?*, Communications of the ACM 34 (1991), pp. 52–64.

An industry advisory board convened in June 1991 in Naples, Florida, to identify the most pressing needs in software support for parallel scientific computing. This article summarizes a two-day discussion by outlining the problems in providing software support to the scientific and technical user community. It includes a description of current and short-term directions in parallel software, and long-range needs in programming languages and environments. It concludes with a presentation of the board's ideas about where software development efforts might best be spent in order to maximize the impact on the high performance user community. Accompanying sidebar pieces provide excerpts from the discussion on the past and future of parallel software.

[15] C. M. Pancake, S. Utter, *A Bibliography of Parallel Debuggers, 1990 Edition*, ACM SIGPLAN Notices 26–1 (1990), pp. 21–37.

The article presents a list of references dealing with parallel debuggers. No distinction is made between tools for *parallel* and for *distributed* systems, although debuggers implemented on multiprocessor or multicomputer architectures which are only capable of analyzing serial programs are omitted. Treatments of parallel programming environments, debugging methodology, or techniques for program analysis are included when a significant portion of the publication is devoted to factors influencing the design or implementation of debugging tools.

Scientific Visualization

[16] K. W. Brodlie, M. Berzins, P. M. Dew, A. Poon, H. Wright, *Visualization and its Use in Scientific Computation*, in "Proceedings of the IFIP Working Conference on Programming Environment for High-Level Scientific Problem Solving", Karlsruhe, 1991, pp. 196–211.

The importance of scientific visualization has increased significantly in recent years for three reasons: The enormous computing power of today's parallel architectures extend the range of numerical simulation experiments, which, in turn, generate vast amounts of data that needs to be converted into color images to be efficiently understood by humans. Advances in data collection equipment technology requires rapid processing. Finally, modern programs, running on interactive, window-based workstations, have brought visualization technology to the scientist's desk.

After an overview of current work in scientific visualization that has provided input and stimulus to the GRASPARC project, the project itself is described and future plans are discussed. The GRASPARC project is an environment for numerical simulation and visualization. A prototype has been realized.

[17] T. A. DeFanti, M. D. Brown, B. H. McCormick, *Visualization: Expanding Scientific and Engineering Research Opportunities*, IEEE Computer 22-8 (1989), pp. 12–25.

The authors define *scientific visualization* as follows:
- a tool for discovery and understanding; and
- a tool for communication and teaching.

They recommend a national initiative on visualisation in scientific computing for short-term and long-term needs of toolmakers and toolusers, and they describe how to manage this.

[18] G. Domik, *The Role of Visualization in Understanding Data*, in "New Results and New Trends in Computer Science" (H. Maurer, Ed.), Lecture Notes in Computer Science, Springer-Verlag, Berlin, 1991, pp. 91–107.

Along with an increase in complexity of computer applications there is an increase in the amount of data presented to the user. Visualization might help the user in understanding and managing the complexity of data. The author identifies visualization components for two-dimensional data and discusses the mapping of the data attributes and judgements (defining the goal of the user, e.g. "categorize", "rank", etc.) into graphical elements characterized by visualization primitives (position, length, colour, shape, etc.).

[19] K. A. Frenkel, *The Art and Science of Visualizing Data*, Communications of the ACM 31 (1988), pp. 111–121.

The author reports on the current state of the art in scientific visualization. In particular he reports his impressions about the SIGGRAPH '87 conference. The panel recommends a new initiative in visualization in scientific computing, to get visualization tools into "the hand and minds" of scientists. Scientists and engineers would team up with visualization researchers in order to solve graphics, image processing, human/computer interface, or representational problems grounded in the needs and methods of an explicit discipline. The expectation is that visualization tools would solve hard, driving problems on the one hand, and computational science would be portable to problems on the other. Proposals would be peer reviewed, and awarded for both facilities and projects at national supercomputer centers and elsewhere. Other agencies of government are encouraged to recognize the value of visualization in scientific computing in their missions and support its development accordingly.

[20] S. N. Gupta, M. Zubair, C. E. Grosch, *Visualization: An Aid to Design and Understand Neural Networks in a Parallel Environment*, in "Proceedings of the Fifth Distributed Memory Computing Conference, Charleston, South Carolina, Vol. 2: Architectures, Software Tools and other General Issues" (Q. Stout, M. Wolfe, Eds.), IEEE Press, Washington DC, 1990, pp. 1223–1227.

Two visualization tools developed on the DAP-510 (a SIMD machine with 1024 processors), are described. First the *interactive visualization tool*, which allows the user to steer the course of computation by interactively modifying its parameters based on visual feedback. Second the *display tool* DT, which transforms the numeric data into a visual form. It also gives the user the capability to manipulate the visual representation. These tools are utilized for designing and understanding the neural networks. But these tools are general in nature and can easily interact with other parallel computation processes.

[21] Y. L. Lous, *Report on the First Eurographics Workshop on Visualization in Scientific Computing*, Computer Graphics Forum 9 (1990), pp. 371–372.

The author summarizes briefly the main aspects and trends presented at the first Eurographics Workshop on *Visualization in Scientific Computing* in France, 1990. 26 papers were presented in 6 sessions:

- The first session was dedicated to the specific needs of the visualization in computational science.
- The second session was devoted to the importance and the difficulties of using standards in visualization software.
- In the third session some reference models and distributed graphics systems were presented.
- In the "Application" session were presented a general 3D visualization system used for postprocessing in various aerodynamic research projects.
- The number of presentations received for the session "Rendering Techniques" emphasized if necessary the importance of global visual representations in visualization graphics.

- In the last session, devoted to data representation and interactions, was demonstrated an effective octree data structure enabling real time animations of 2D color contour maps on a PC/AT with VGA card.

[22] B. H. McCormick, T. A. DeFanti, M. D. Brown, *Visualization in Scientific Computing*, IEEE Computer Graphics 21 (1987), pp. 1–14.

Visualization in Scientific Computing is emerging as a major computer-based field, with a body of problems, a communality of tools and terminology, boundaries, and a cohort of trained personnel. As a tool for applying computers to science, it offers a way to see the unseen. As a technology, Visualization in Scientific Computing promises radical improvements in the human/computer interface and may make human-in-the-loop problems approachable.

Visualization in Scientific Computing can bring enormous leverage to bear on scientific productivity and the potential for major scientific breakthroughs, at a level of influence comparable to that of supercomputers themselves. It can bring advanced methods into technologically intensive industries and promote the effectiveness of the American scientific and engineering communities. Major advances in Visualization in Scientific Computing and effective national diffusion of its technologies will encourage the development of techniques for understanding how models evolve computationally, for tightening design cycles, for integrating hardware and software tools, and for standardizing user-interfaces.

[23] C. Mundie, *Tools for Scientific Visualization*, in "Digest of Papers (COMPCON)" (H. S. AlKhalib, K. Miura, Eds.), IEEE Press, San Francisco, 1989, pp. 320–321.

Possibilities to realize *scientific visualization* are shown. Modern scientists need a mini-supercomputer or supercomputer class server with integrated graphics, networked to PC and workstations on the desktop. The article presents a chemistry example to show how such a server environment works.

[24] G. M. Nielson, *Visualization in Scientific Computing*, IEEE Computer 22-8 (1989), pp. 10–11.

Scientific visualization techniques in use today are discussed and new research developments in this area are presented.

[25] C. Upson, *Scientific Visualization Environments for the Computational Sciences*, in "Digest of Papers (COMPCON)" (H. S. AlKhalib, K. Miura, Eds.), IEEE Press, 1989, pp. 322–327.

The process of *scientific visualization* is presented. It describes the transformations necessary from observing nature to simulating it (observations, physical laws, mathematical formulations of laws, simulation specification, simulation solution, analysis of images). The problem is, that scientists can formulate equations and compute their solutions faster than they can analyze their meaning. Due to this fact scientists are looking to visualization for help. The article describes the analysis cycle. It consists of three data transformational stages and an image playback stage.

Program Visualization

[26] M. Abrams, N. Doraswamy, A. Mathur, CHITRA: *Visual Analysis of Parallel and Distributed Programs in the Time, Event, and Frequency Domains*, IEEE Transactions on Parallel and Distributed Systems 3 (1992), pp. 672–685.

> CHITRA analyzes a program execution sequence (PES) collected during execution of a program and produces a homogeneous, semi-Markov chain model fitting the PES. The PES represents the evolution of a program state vector in time. Therefore CHITRA analyzes the time-dependent behavior of a program. The article describes a set of transformations that map a PES to a simplified PES. As the transformations are program-independent, CHITRA can be used with any program. CHITRA provides a visualization of PES and transformations, to allow a user to visually guided transformation selection in an effort to generate a simple yet accurate semi-Markov chain model. The resultant chain can predict performance at program parameters different from those used in the input PES, and the chain structure can diagnose performance problems.

[27] S. Baker, H.-J. Beier, T. Bemmerl, A. Bode, H. Ertl, U. Graf, O. Hansen, J. Haunerdinger, P. Hofstetter, R. Knödlseder, J. Kremenek, S. Langenbuch, R. Lindhof, T. Ludwig, P. Luksch, R. Milner, B. Ries, T. Tremi, TOPSYS - *Tools for Parallel Systems*, SFB Report 342/13/91 A, Institute for Computer Science, Technical University of Munich, 1991.

> TOPSYS is an integrated environment for programming distributed memory multiprocessors. Its main goal is to hide the architecture of the hardware from programmers and make a first step towards virtualizing the hardware of scaleable multiprocessor systems. TOPSYS offers tools for performance analysis (PATOP), visualization (VISTOP), specification and mapping (SAMTOP), and a source oriented parallel debugger (DETOP). The basis for all these tools is the MMK (Multiprocessor Multitasking Kernel) which supports dynamic load balancing. Since TOPSYS is a hierarchical system it supports portability to new multiprocessor systems, and offers expandability to new tools. All tools use a common graphical interface based on the X window system and are integrated in order to reduce the implementation time of the entire environment.

[28] T. Bemmerl, *Programming Tools for Massively Parallel Supercomputers*, in "Environments and Tools for Parallel Scientific Computing" (J. J. Dongarra, B. Tourancheau, Eds.), Elsevier Science Publishers, Amsterdam, 1993, pp. 125–136.

> This paper covers programming tools activities going on within Intel Supercomputers in general and the European Supercomputer Development Center in particular. These activities target toward a smooth transition of the parallelization methodology from manual to interactive (semiautomatic) and finally to fully automated. Integrated programming tool environments will support the application of this parallelization methodology. Most attention will be given within this paper to parallel program tuning tools for massively parallel architectures, like the Paragon XP/S family of supercomputers.

[29] D. Bernstein, A. Bolmarcich, K. So, *Performance Visualization of Parallel Programs on a Shared Memory Multiprocessor System*, in "International Conference on Parallel Processing" (E. C. Plachy, Ed.), Pennsylvania State University Press, Los Angeles, 1989, pp. 1–10.

PREFACE is a *performance monitoring tool* for visualizing the execution of parallel programs on a shared memory multiprocessor system. The performance data can be visualized during the execution of the program or they can be recorded in event traces and displayed to the user after the parallel program has terminated. The parallel program is instrumented by a preprocessor. Three parameters are measured for each chore[3]:

- the total execution time;
- the execution time spent in the user space; and
- the execution time spent in the system space.

The tool displays the utilization of the participating processors according to where their execution time is spent (for instance: user and system work, parallelization overhead, idle time).

[30] H. Blaschek, G. Drsticka, G. Kotsis, *Visualisation of Parallel Program Behavior*, Report, Institute for Statistics and Computer Science, University Vienna, 1992.

This paper is a survey of the state of the art in behavior and performance visualization in parallel processing. The pictorial representation of performance data and the use of metaphors for hardware and software components will help the programmer or system analyst to obtain a better understanding of what is going on in the program, how it works, and why it works. A number of visualization tools are available so far; some of them are mainly tools for algorithm animation, others are aimed to support debugging or focus on visualization of performance data. The authors enlist relevant work published in this field and make suggestions on how a visualization tool might look, what metaphors and display types are suitable for representing different aspects of a parallel system, including both hardware and software characteristics.

[31] A. Bode, P. Braun, *Monitoring and Visualization in* TOPSYS, in "Performance Measurement and Visualization of Parallel Systems" (G. Haring, G. Kotsis, Eds.), Elsevier Science Publishers, Amsterdam, 1993, pp. 97–129.

The paper describes the on-line visualization and *animation system* VISTOP (VISualization TOol for Parallel systems) and its use of the TOPSYS distributed monitoring system. TOPSYS (TOols for Parallel SYStems) is an integrated tool environment which aims at simplifying the usage and programming of parallel systems. It consists of several interactive development tools for specification, debugging, performance analysis and visualization. VISTOP is presented after an evaluation and classification of existing visualization environments. It supports the interactive on-line visualization of message passing programs based on various views, in particular,

---

[3]A *chore* is defined as the amount of work assigned to an executing element for execution at a time.

a process graph based concurrency view for detecting synchronization and communication errors. The data acquisition of VISTOP is performed using distributed monitoring system common to all tools of the TOPSYS environment.

[32] O. Brewer, J. J. Dongarra, D. Sorensen, *Tools to aid in the analysis of memory access patterns for Fortran programs*, Parallel Computing 9 (1988), pp. 25–35.

The efficiency of algorithms developed for high performance computers is not only characterized by the number of memory accesses but also the access patterns. The authors present a set of tools designed to aid in the analysis of *memory access patterns* of Fortran programs. The MAPI (Memory Access Pattern Instrumentation) tool instruments a user's program and produces a trace file when running the instrumented program. This trace file serves as an input for the MAPA (Memory Access Pattern Animation) tool, which displays read and write accesses to array cells by displaying the array (at most 4 at a time) and by highlighting the cells (read access as blue flash and write access as a red flash).

[33] M. H. Brown, *Algorithm Animation*, MIT Press, Cambridge MA, 1988.

[34] M. H. Brown, *Exploring Algorithms Using* BALSA-II, IEEE Computer 21-5 (1988), pp. 14–36.

BALSA-II is an "exploratorium" for investigating the *dynamic behavior* of programs. Users watch and interact with the simulations from various views. Each view is displayed in a window. Programmers implement the algorithms, input generators and views that the users see and manipulate. The article is organized in two major sections: the first part looks at BALSA-II from the user's perspective, the second from the programmer's.

[35] M. H. Brown, ZEUS: *A System for Algorithm Animation and Multi-View Editing*, in "Proceedings of the 1991 IEEE Workshop on Visual Languages" (D. P. Stotts, Ed.), IEEE Press, Los Alamitos CA, 1991, pp. 4–9.

ZEUS, an algorithm animation system, can automatically generate some views based on the events an algorithm generates. Each view is an animated picture portraying the events as they are generated by the algorithm.

[36] M. H. Brown, J. Hershberger, *Color and Sound in Algorithm Animation*, in "Proceedings of the 1991 IEEE Workshop on Visual Languages" (D. P. Stotts, Ed.), IEEE Press, Los Alamitos CA, 1991, pp. 10–17.

The techniques for algorithm animation that are reported in the literature are reviewed and new techniques for using sound and color are introduced.

[37] M. H. Brown, J. Hershberger, *Color and Sound in Algorithm Animation*, IEEE Computer 25-12 (1992), pp. 52–63.

This article is more or less similar to Brown, Hershberger [36], but more examples are given and the text has been expanded.

[38] M. H. Brown, R. Sedgewick, *A System for Algorithm Animation*, IEEE Computer Graphics 18-3 (1984), pp. 177–185.

The software environment BALSA, which provides facilities at a variety of levels for "animating" algorithms, is described. BALSA exposes properties of programs by displaying multiple dynamic views of a program and associated data structures. The system is operational on a network of graphics-based, personal workstations and has been used successfully in several applications for teaching and research in computer science and mathematics. In this paper, the authors outline the conceptual framework that they have developed for animating algorithms, describe the system that they have implemented, and give several examples drawn from the host of algorithms that they have animated.

[39] M. H. Brown, R. Sedgewick, *Techniques for Algorithm Animation*, IEEE Software 2-1 (1985), pp. 28–39.

BALSA (Brown ALgorithm Simulator and Animator) is a tailored special-purpose system for the *animation* of sequential algorithms providing dynamic (real-time) graphic displays. Application fields are science education, design and analysis of algorithms, advanced debugging and system programming. When preparing an algorithm for animation, event signals must be added indicating interesting points in the program's execution which shall lead to changes in the display. The "animator" writes the corresponding software for graphic displays. The resulting set of algorithms and views can either be used by the user directly (invoking the BALSA interpreter and window manager) or by invoking scripts (predefined sequences of operations created by a "scriptwriter").

[40] H. Chernoff, *The use of faces to represent points in k-dimensional space graphically*, Journal of the American Statistical Association 68 (1973), pp. 361–368.

A method of representing multivariant data is presented. Each point of a k-dimensional space, $k \leq 18$, is represented by a cartoon of a face whose features, such as length of nose and curvature of mouth, correspond to components of the point. Thus every multivariant observation is visualized as a computer-drawn face. This presentation makes it easy for the human mind to grasp many of the essential regularities and irregularities present in the data.

[41] A. L. Couch, *Graphical representations of program performance*, Technical Report TR 88-4, Department of Computer Science, Tufts University, Medford MA, 1988.

[42] H. Davis, S. R. Goldschmidt, J. Hennessy, *Multiprocessor Simulation and Tracing Using* TANGO, in "International Conference on Parallel Processing, Vol. 2" (H. D. Schwetman, Ed.), Pennsylvania State University Press, Los Angeles, 1991, pp. 99–107.

TANGO is a software *simulation* and *tracing* system used to obtain data for evaluating parallel programs and multiprocessor systems. The system provides a simulated multiprocessor environment by multiplexing application processes onto a single processor. So software running on an available host machine is used to emulate the program execution on some target machine. Two alternative simulation techniques are discussed: trace-driven and execution-driven simulation. The computing environment and the implementation of TANGO and its performance are described. Sources of simulation overhead are discussed.

[43] I. M. Demeure, G. J. Nutt, *The* VISA *Distributed Computation Modelling System*, in "Proceedings of the 5th International Conference on Modelling Techniques and Tools for Computer Performance Evaluation" (N. Abu El Ata, Ed.), Elsevier Science Publishers, Torino, Italy, 1991, pp. 133–147.

When developing parallel and distributed algorithms, partitioning the computation into processes and designing efficient interprocess communication strategies are challenging tasks. VISA is an environment for *modelling* and *simulating* different partitioning and communication strategies for large-grained MIMD computations. The parallel, distributed computation graph model (ParaDiGM), a formal graph model, is used as the basis for VISA's operation. The user can create, edit and simulate ParaDiGM models. A qualitative reporter serves as an *animation* facility, and a quantitative reporter displays simulation results (using graphs and tables) either during simulation or summarized at the end of a simulation run.

[44] J. J. Dongarra, O. Brewer, J. A. Kohl, S. Fineberg, *A Tool to Aid the Design, Implementation, and Understanding of Matrix Algorithms for Parallel Processors*, Journal of Parallel and Distributed Computing 9 (1990), pp. 185–202.

[45] J. J. Dongarra, D. Sorensen, O. Brewer, *Tools to aid in the design, implementation, and understanding of algorithms for parallel processors*, in "Software for Parallel Computers" (R. H. Perrott, Ed.), Chapman & Hall, London, 1992, pp. 195–219.

The authors have developed two tools that aid in the development of parallel algorithms and are portable across a range of high-performance computers. The first tool, SCHEDULE, aids in implementing and analyzing programs within a large-grain control flow model of computation. The underlying concept is based on a natural graphical interpretation of parallel computation that is useful in designing and implementing parallel algorithms. This graphical interpretation may be used to automate the generation of a parallel program through a facility called BUILD. SCHEDULE also provides means for postprocessing performance analysis through an animated visualization of the flow of a parallel program's execution. This animation is accomplished through the TRACE facility. The second tool provides a graphical display of memory access patterns in algorithms and is described in detail in Dongarra et al. [44].

[46] R. Earnshaw, N. Wiseman, *An Introductory Guide to Scientific Visualization*, Springer-Verlag, Berlin Heidelberg New York Tokyo, 1992.

[47] H. El-Rewini, T. G. Lewis, *Scheduling Parallel Program Tasks onto Arbitrary Target Machines*, Journal of Parallel and Distributed Computing 9 (1990), pp. 138–153.

TaskGrapher is a tool for studying optimum *parallel program task scheduling* on (arbitrarily interconnected) processors. Once the parallel program (represented as a task graph) along with the interconnection topology of the target machine (drawn in a topology window) is entered, the user can apply several heuristics for scheduling. TaskGrapher provides the following displays:

- Gantt Chart Scheduler;
- Speedup Line Graph;
- Critical Path Display of the Task Graph;
- Processor Utilization Chart;
- Processor Efficiency Chart; and
- Dynamic Activity Display.

[48] D. E. Foulser, W. D. Gropp, CLAM and CLAMSHELL: *An Interactive Front-End for Parallel Computing and Visualization*, in "International Conference on Parallel Processing, Vol. 3" (P.-C. Yew, Ed.), Pennsylvania State University Press, Los Angeles, 1990, pp. 35–43.

CLAM, a Computational Linear Algebra Machine, and CLAMSHELL, an interactive computing environment and interface, are presented. CLAM and CLAMSHELL make use of X Window Systems graphics. CLAMSHELL graphics include 2D plots, 3D line drawing, surface plots, contour plots and volume rendering. Any image can be drawn statically or animated.

[49] R. J. Fowler, T. J. LeBlanc, J. M. Mellor-Crummey, *An Integrated Approach to Parallel Program Debugging and Performance Analysis on Large-Scale Multiprocessors*, ACM SIGPLAN Notices 24–1 (1989), pp. 163–173.

A toolkit for *parallel program debugging and performance analysis* is described. The toolkit is used in a two-phase-model of program analysis, in which a program's execution is monitored, with the resulting data forming the basis for later, off-line analysis. The execution history browser, called MOVIOLA, implements a graphical view of an execution based on the directed acyclic graph representation of processes and communication. The capabilities of this tool are presented on the algorithm for odd even merge sort on a *butterfly network*.

[50] J. M. Francioni, J. A. Jackson, L. Albright, *The Sounds of Parallel Programs*, in "Proceedings of the Sixth Distributed Memory Computing Conference, Portland, Oregon" (Q. Stout, M. Wolfe, Eds.), IEEE Press, Washington DC, 1991, pp. 570–577.

In order to understand the behavior of parallel programs *auralization* is proposed as an alternative to visualization.

[51] R. Friedhoff, W. Benzon, *Visualization: The Second Computer Revolution*, Freeman, New York, 1991.

[52] C. A. Funka-Lea, T. D. Kontogiorgos, R. J. T. Morris, L. D. Rubin, *Interactive Visual Modelling for Performance*, IEEE Software 8–5 (1991), pp. 58–68.

The Q+ simulator is a discrete-event simulation system for queuing network models. Using a graphical editor, queuing network models can be constructed and parametrized in a convenient way. The user has the possibility to observe the movement of the customers through the network (animation during simulation) and to obtain post-mortem statistics for service centers and for the whole system.

[53] I. Glendinning, V.S. Getov, S.A. Hellberg, R.W. Hockney, D.J. Pritchard, *Performance Visualization in a Portable Parallel Programming Environment*, in "Performance Measurement and Visualization of Parallel Systems" (G. Haring, G. Kotsis, Eds.), Elsevier Science Publishers, Amsterdam, 1993, pp. 251–275.

In order to obtain the highest possible performance from programs running on massively parallel machines it is essential to identify precisely where and when computational resources are consumed during the execution. A number of performance visualization tools have evolved to meet this need for particular systems but they are often not portable to other machines. The authors regard portability as crucial to the widespread acceptance and use of such tools, and they have investigated several approaches to achieve it. Each approach has been based on the public domain PARAGRAPH tool, which enables trace data collected during a program's execution to be viewed from various different visual perspectives. One approach is for programs to use the portable instrumented communication library PICL, which directly generates trace data in the appropriate format. Alternatively, trace files produced by applications using other libraries can be converted into PARAGRAPH format using trace filter programs. In this paper the authors report on an implementation of PICL for transputers and on trace filters developed for the PARMACS and EXPRESS libraries. The authors also describe ongoing work within the PPPE Esprit project to integrate PARAGRAPH into a portable parallel programming environment based on the PCTE portable common tool environment.

[54] V.A. Guarna, D. Gannon, D. Jablonowski, A.D. Malony, Y. Gaur, FAUST: *An Integrated Environment for Parallel Programming*, IEEE Software 6–4 (1989), pp. 20–27.

FAUST is an integrated *environment for parallel programming*. It consists of four major parts:

1. the project manager for organizing and manipulating the project;

2. the project database including 8 types of files;

3. FAUST tools (e.g. editors, compilers, etc.); and

4. FAUST building blocks (e.g. graph manager/browser).

FAUST includes the following user-level tools:

1. SIGMA, designed to help users of parallel supercomputers in retargeting and optimizing application code;

2. CALL-GRAPH, a tool showing an animated view of a program's execution to help users understand its operation; and

3. IMPACT, an event-display tool tracing multitasking events and displaying them in a time line.

[55] G. Hansen, C. Linthicum, G. Brooks, *Experience with a Performance Analyser for Multithread Applications*, Proceedings of Supercomputing '90 (1990), pp. 124–131.

Determining the effectiveness of parallelization requires performance data about elapsed time and total CPU time. Furthermore, it is desirable not to have to run a parallel application in a stand-alone environment in order to obtain a profile. The paper describes the Convex performance analyzer, CXPA, which has the capability to monitor parallel regions of code, in particular loops, executing in a time-sharing environment. The means by which profiling information is measured for a parallel region is described along with the operating system facilities required to support it. The effectiveness of the approach is evaluated and suggestions for improvements are made.

[56] M. T. Heath, *Visual Animation of Parallel Algorithms for Matrix Computations*, in "Proceedings of the Fifth Distributed Memory Computing Conference, Charleston, South Carolina, Vol. 2: Architectures, Software Tools and other General Issues" (Q. Stout, M. Wolfe, Eds.), IEEE Press, Washington DC, 1990, pp. 1213–1222.

In this paper the author shows how graphical animation of the behavior of parallel algorithms can facilitate the design and performance enhancements of algorithms for matrix computations on parallel computer architectures. Using the portable instrumented communication library PICL (Geist et al. [4]) and the graphical animation package PARAGRAPH (Heath, Etheridge [57]) the author illustrates the effects of various strategies in parallel algorithm design, including interconnection topologies, global communication patterns, data mapping schemes, load balancing, and pipelining techniques for overlapping communication with computation. This article focuses on distributed memory parallel architectures in which the processors communicate by passing messages. The linear algebra problems which are considered include matrix factorization and the solution of triangular systems.

[57] M. T. Heath, J. A. Etheridge, *Visualizing Performance of Parallel Programs*, Technical Report ORNL/TM–11813, Oak Ridge National Laboratory, 1991.

In this paper the authors describe a graphical display system, called PARAGRAPH, for visualizing the behavior and performance of parallel programs on message-passing multiprocessor architectures. The visual animation is based on executing trace information monitored during an actual run of a parallel program on a message-passing parallel computer. The resulting trace data is replayed pictorially to provide a dynamic depiction of the behavior of the parallel program, as well as graphical summaries of its overall performance. Several distinct visual perspectives are provided from which to view the same performance data, in attempt to gain insights that might be missed by any single view. The authors describe this visualization tool, outline the motivation and philosophy behind its design, and illustrate its usefulness in analyzing parallel programs.

[58] M. T. Heath, J. A. Etheridge, *Visualizing the Performance of Parallel Programs*, IEEE Software 8-5 (1991), pp. 29–39.

PARAGRAPH is a graphical display tool for *visualizing parallel programs* run on message passing multiprocessor architectures based on trace data generated by PICL (Geist et al. [4]). It offers the user several possibilities for replaying the programs dynamic behavior as well as graphical summaries of its overall performance. PARAGRAPH offers utilization displays (count, Gantt Chart, summary, meter, profile, Kiviat diagram), communication displays (traffic, space-time diagram, message queues, communication matrix, meter, animation, node statistics), and task displays (count, Gantt Chart, status, summary). Further on, the user can add application specific displays by extending PARAGRAPH (linking the necessary routines written by the user). A detailed discussion of PARAGRAPH can be found in Heath, Etheridge [57].

[59] M. T. Heath, *Recent Developments and Case Studies in Performance Visualization using* PARAGRAPH, in "Performance Measurement and Visualization of Parallel Systems" (G. Haring, G. Kotsis, Eds.), Elsevier Science Publishers, Amsterdam, 1993, pp. 175–200.

PARAGRAPH is a graphical display system for visualizing the behavior and performance of parallel programs on message-passing multicomputer architectures. In this paper, the author discusses a number of new features recently added to PARAGRAPH, and he illustrates its usefulness by means of several case studies of performance tuning based on visual feedback.

[60] E. Helttula, A. Hyrskykari, K.-J. Räihä, *Graphical Specification of Algorithm Animations with* ALADDIN, in "Proceeding of the Hawaii International Conference on System Sciences, Vol. II" (B. D. Shriver, Ed.), IEEE Computer Society Press, Washington DC, 1989, pp. 892–901.

ALADDIN (ALgorithm Animation Design and Description using INteraction) is a system for creating animations of algorithm executions with minimal effort. To avoid the most laborious phase of producing an animation, which is programming the graphics needed in the animation, ALADDIN allows the user to design the graphics interactively with a graphical editor. An example elucidating how an animation can be designed using the system is given. Some details of the implementation of the system are examined more closely.

[61] A. A. Hough, J. E. Cuny, BELVEDERE: *Prototype of a Pattern-Oriented Debugger for Highly Parallel Computation*, in "International Conference on Parallel Processing" (S. K. Sahni, Ed.), Pennsylvania State University Press, Los Angeles, 1987, pp. 735–738.

Often the behavior of a highly parallel software system is best understood in terms of its patterns of interprocess control and data flow. The authors describe the design of a pattern-oriented debugger, called BELVEDERE, which provides facilities for the animation and manipulation of these patterns of interactions. BELVEDERE automatically animates primitive system events and abstract, user-defined events. In addition, it provides facilities for restricting the display according to a variety of criteria and for viewing system behavior from user-defined perspectives.

[62] A. A. Hough, J. E. Cuny, *Initial Experiences with a Pattern-Oriented Parallel Debugger*, ACM SIGPLAN Notices 24-1 (1989), pp. 195–205.

BELVEDERE facilitates the description, manipulation, and *animation* of logically structured patterns of process interaction. In the animation, processes are shown as large squares, ports as small squares, and channels as connecting lines. Animated events are highlighted. Because in some cases the expected "checker board" patterns are not evident, BELVEDERE allows the user to define abstract events. These events are inserted into the event stream where they become available as the target of queries. BELVEDERE uses two types of perspectives to restrict the displayed behavior:

• the process perspective, i.e. a view seen from a single process; and
• the consensus perspective, animating high level events in a sequence consistent with that seen by all participating processes.

[63] A. A. Hough, J. E. Cuny, *Perspective Views: A Technique for Enhancing Parallel Program Visualization*, in "International Conference on Parallel Processing, Vol. 2" (D. A. Padua, Ed.), Pennsylvania State University Press, Los Angeles, 1990, pp. 124–132.

Techniques for "reordering" events to enhance *visualization of abstract events* are described. Abstract nonatomic events may overlap in both time and space. Consistent reordering preserves the partial ordering of events imposed by the sequentiality of processes and by the interprocess dependencies. For many parallel computations reorderings are not possible because of interwined dependencies. For these cases the concept of *perspective views* is introduced. Partially consistent reorderings are established, which provide a partial view of the systems behavior. Several partial views may be needed to understand its full behavior. Perspective views enhance visualization by reordering behavior so that abstract events can be seen as distinct units. The algorithm for computing perspective views can be applied to any stream of locally time stamped events and any display system that produces a visualization of the recorded behavior.

[64] K. Imre, *Experiences with Monitoring and Visualization the Performance of Parallel Programs*, in "Performance Measurement and Visualization of Parallel Systems" (G. Haring, G. Kotsis, Eds.), Elsevier Science Publishers, Amsterdam, 1993, pp. 19–44.

Tuning the performance of parallel programs is a crucial but difficult task which requires a good understanding of the program to be tuned. The aim of performance monitoring and visualization tools is to give this good understanding to a programmer. Any parallel computation that has large number of processes makes most of the visualization techniques obsolete since the volume of performance data to be displayed is much higher than the volume of information that a human observer can comprehend. To extract useful information from large volumes of performance data and to visualize these data in a proper form, new techniques are required. In this paper, the author presents a techniques to extract useful information from large volumes of performance data, and to visualize this information in several different graphical forms. Since the user can assign semantic knowledge to the information to be visualized, it is easy and very flexible to create abstract views which can be interpreted in the context of application-specific or case-sensitive performance visualization.

[65] S. Isoda, T. Shimomura, Y. Ono, VIPS: *A Visual Debugger*, IEEE Software 4-3 (1987), pp. 8-19.

VIPS uses graphics to *show the static and dynamic behavior* of an Ada program execution. The compiler of Ada generates two files: the Diana file containing the syntax tree of the Ada program, and the quadruple file containing a sequence of tuples of an operation and its operands. The preprocessor of VIPS analyses the Diana file for information about blocks and variables. The several views of the Ada program execution are represented in windows: data and figure definition windows (displaying a list of variables and the name of user-defined figures), an editor and a program text window (for editing and displaying the source program), a block structure window (showing the nesting relationship of subprograms and internal packages), an acceleration window (displaying the current execution speed), and an interaction window (displaying the interaction between a user and a test program). For the presentation standard figures and figures, defined by the user in the Figure Description Language, can be used.

[66] R. Khanna, B. McMillin, *A Visualization Model for Massively Parallel Algorithms*, in "Proceedings of the Sixth Distributed Memory Computing Conference, Portland, Oregon" (Q. Stout, M. Wolfe, Eds.), IEEE Press, Washington DC, 1991, pp. 617-620.

SMILI is a visualization tool mainly for the representation of multivariant data by means of Chernoff Faces.

[67] D. Kimelman, G. Sang'udi, *Program Visualization by Integration of Advanced Compiler Technology with Configurable Views*, in "Environments and Tools for Parallel Scientific Computing" (J. J. Dongarra, B. Tourancheau, Eds.), Elsevier Science Publishers, Amsterdam, 1993, pp. 73-84.

A major challenge facing program visualization is to provide more effective displays of program behavior. This paper proposes language-level program structure, as derived by advanced compiler technology, as a basis for more effective displays. The experimental integration of a program visualization environment with a parallelizing compilation system, undertaken in order to achieve such displays, is described.

Programs are parallelized and compiled, and then executed on a parallel system. Program source, static call graphs, control and data flow graphs, and control dependence graphs are produced during compilation, and then presented to the user by the visualization system. These displays are then animated based on a trace containing an execution history produced by the language run time system. The resulting displays will allow more effective presentation of program behavior for parallel and distributed systems.

[68] K. Kolence, P. Kiviat, *Software Unit Profiles and Kiviat Figures*, ACM SIGMETRICS, Performance Evaluation Review 2-3 (1973), pp. 2-12.

[69] M. S. Krishnamoorthy, R. Swaminathan, *Program Tools for Algorithm Animation*, Software Practice and Experience 19 (1989), pp. 505-513.

Algorithm animation is a technique used in computer aided instruction. The effectiveness of its usage depends on how easily the animation tools can be handled. The authors propose graphics primitives for constructing programs for algorithm animation. Those primitives help in storing the parameters of objects, in moving objects and so on, thus reducing the time for animation program development.

[70] M. V. LaPolla, *Toward a Theory of Abstractions and Visualizations for Massively Parallel Computing*, Technical Report Org. 96–40, Lockheed Advanced Computation Laboratory, Palo Alto CA, 1991.

Programming and debugging software for massively parallel machines is still very much like conventional programming in the late 50's and early 60's. The work is highly machine-dependent; the province of small groups of adepts specializing in particular applications and hardware configurations. Most of the debugging (and programming) is done by these "wizards" by trial and error. Trial and error is used partly because there are no tools for apprehending and interpreting the large amounts of data computed by these machines; therefore, it is necessary to invent tools. The tools needed include:

1. data abstraction for representing the salient information in massively parallel data structures and the operations on these data structures;

2. visualizations for apprehending the information encoded in the abstract data structures and the effects of operations on them; and

3. a method for interpreting visualizations and applying the knowledge gained from interpretation to debugging the program that produced the effect.

[71] M. V. LaPolla, *The View of Parallel Programming*, in "Two-day Workshop on Parallel Programming Tools '91: Twenty-fifth Hawaii International Conference on System Sciences", Computer Society Press, Los Alamitos CA, 1991.

IVE (Integrated Visualization Environment) allows a user to create *visualizations* of the state of a *massively parallel machine*. It supports three different kinds of visualization:

- *program visualizations* (presenting logical and structural software relationships);

- *process visualizations* (depicting the execution of a program); and

- *application visualizations* (representing the results of computations using algorithm specific abstractions).

[72] T. J. LeBlanc, J. M. Mellor-Crummey, R. J. Fowler, *Analyzing Parallel Program Executions Using Multiple Views*, Journal of Parallel and Distributed Computing 9 (1990), pp. 203–217.

The authors propose *parallel program analysis* based of a *multiplicity of views* on an execution. The framework for analysis is defined in terms of the information to be analysed. Analysis follows the development cycle (from error detection and diagnosis to performance analysis and tuning) and proceeds top-down (from abstract views to more specific ones). The program's behavior is represented as a graph, from which the user can generate fine-tune visualizations using an integrated programmable toolkit. The views are characterized by three dimensions, process interaction, process state and time. This characterization is also used to classify related work.

[73] T. Lehr, D. Black, Z. Segall, D. Vrsalovic, *Visualizing Context-Switches Using* PIE *and the* MACH *Kernel Monitor*, in "International Conference on Parallel Processing, Vol. 2" (D. A. Padua, Ed.), Pennsylvania State University Press, Los Angeles, 1990, pp. 298–299.

The MACH scheduler uses the MKM, the MACH context-switch monitor, and the PIE programming environment (Lehr et al. [75]). The MKM monitors context-switches of selected program threads. The context-switching patterns of the program are then displayed by PIE using a number of *visual displays* (execution barscope view and CPU barscope view). The user has the possibility to distinguish between the program and scheduler performance problems.

[74] T. Lehr, D. Black, Z. Segall, D. Vrsalovic, *Visualizing System Behavior*, in "International Conference on Parallel Processing, Vol. 2" (H. D. Schwetman, Ed.), Pennsylvania State University Press, Los Angeles, 1991, pp. 117–123.

A method for analyzing the effect of operating system behavior upon program *performance* is presented. The MACH Kernel Monitor is built into MACH 2.5, an operating system. It tracks context-switches of selected threads. The graphical interface to the monitor data was developed as a part of the Parallel Programming and Instrumentation Environment PIE (Lehr et al. [75]). The context-switching patterns of a program are mapped onto an execution graph which is displayed using a number of visual formats (PIE execution barscope view).

[75] T. Lehr, Z. Segall, D. F. Vrsalovic, E. Caplan, A. L. Chung, C. E. Fineman, *Visualizing Performance Debugging*, IEEE Computer 22-10 (1989), pp. 38–51.

A special *software development environment* called the Parallel Programming and Instrumentation Environment (PIE) is examined. PIE is designed to develop performance efficient parallel and sequential programs. PIE supports a software development methodology extended to the analysis, verification and validation of a computation's performance. The PIE system helps the user in predicting, detecting and avoiding performance degradation. When running a program, PIE gathers trace information via a software monitor. The trace data are displayed using Gantt Charts. PIE supports several programming languages (C, Ada, Fortran, etc.). It uses a pictorial representation of programming constructs.

[76] R. L. London, R. A. Duisberg, *Animating Programs Using Smalltalk*, IEEE Computer 18-8 (1985), pp. 61–71.

Algorithm animation plays an important role in designing and debugging programs. Visualization may help the user in understanding how and why a program will work or not. The authors present an object-oriented approach for a visualization programming environment using the Model-View-Controller (MVC) of Smalltalk. The MVC turned out to be a well suited tool for creating animated views of an algorithm.

[77] A. D. Malony, G. V. Wilson, *Future Directions in Parallel Performance Environments*, in "Performance Measurement and Visualization of Parallel Systems" (G. Haring, G. Kotsis, Eds.), Elsevier Science Publishers, Amsterdam, 1993, pp. 331–351.

The increasing complexity of parallel computing systems has brought about a crisis in parallel performance evaluation and tuning . Although there have been important advances in performance tools in recent years, the authors believe that future parallel performance environments will move beyond these tools by integrating performance instrumentation with compilers for architecture-independent languages, by formalizing the relationship between performance views and the data they represent, and by automating some aspects of performance interpretation. The paper describes these directions from the perspective of research projects that have been recently undertaken.

[78] A. D. Malony, JED – *Just an Event Display*, in "Parallel Computer Systems: Performance Instrumentation and Visualization" (M. Simmons, R. Koskela, I. Bucher, Eds.), ACM Press, New York, 1990.

[79] A. D. Malony, *Performance Observability*, Ph.D. Thesis, Department of Computer Science, University of Illinois at Urbana-Champaign, 1990.

In Chapter 6 of Malony's thesis, several examples of *performance visualization techniques and tools* are discussed. HYPERVIEW, a performance visualization environment, is presented, designed to build a framework for the future development and integration of new tools. Performance data are collected in a "database" from which views can be selected. The performance visualizer binds the views dynamically to the selected types of displays. This concept should lead to modularity and extendibility.

[80] A. D. Malony, D. H. Hammerslag, D. J. Jablonowski, TRACEVIEW: *A Trace Visualization Tool*, IEEE Software 8-5 (1991), pp. 19–28.

TRACEVIEW is a *general-purpose trace-visualization tool* where the user can specify a set of trace files (trace manager), define a set of views (view manager), and create a set of displays (display manager) for each view which defines a subregion by setting a beginning and an ending point or by event filtering. The session manager is responsible for the overall coordination; especially for storing and retrieving selected trace files with corresponding views and displays. TRACEVIEW supports multiple simultaneous displays, to compare data from several trace files. The available displays include Gantt Charts (based on the state transitions) and rate displays (which display the number of times a state is entered).

[81] A. D. Malony, D. A. Reed, *Visualizing Parallel Computer System Performance*, Technical Report UIUC-DCS-R-88-1465, Center for Supercomputing Research and Development, University of Illinois at Urbana-Champaign, 1988.

[82] M. A. Marsan, G. Balbo, G. Conte, *Performance Models of Multiprocessor Systems*, MIT Press, Cambridge MA, 1986.

[83] B. Melamed, R. J. T. Morris, *Visual Simulation: The Performance Analysis Workstation*, IEEE Computer 18-8 (1985), pp. 87–94.

The Performance Analysis Workstation, provides an effective *programming tool*, the visual simulation. The system under study is modelled as a queuing network. The user draws the network topology and enters numerical parameters, instruments the simulation and defines statistics, and proceeds to debug and verify the model by tracing its operation through animated simulation runs. The runs may be interrupted and the model modified and reset or resumed interactively.

[84] B. P. Miller, DPM: *A Measurement System for Distributed Programs*, IEEE Transactions on Computers 37 (1988), pp. 243–248.

DPM is a measurement system for *monitoring* the execution and performance of *distributed programs*. It can be used for post-mortem analysis of a program's performance, real-time performance monitoring, and generating data used by the operating system to feedback scheduling activities. The measurement tool consists of four components:

- the meter interface;
- the filter interface;
- the analysis interface; and
- the user interface.

The detection of events is referred to as metering. A trace is produced for each event that is detected. Filtering is the data selection, it reduces the size and the number of traces that are produced. Three analysis are implemented in DPM: basic communication statistics, detecting paths of causality and measuring parallelism in a computation. Several types of *graphic displays* are available for the communication statistics analysis. The "hot spot", matrix and history displays.

[85] B. P. Miller, M. Clark, J. Hollingsworth, S. Kierstead, S.-S. Lim, T. Torzewski, IPS-2: *The Second Generation of a Parallel Program Measurement System*, IEEE Transactions on Parallel and Distributed Systems 1 (1990), pp. 206–217.

IPS is a *performance measurement system* for parallel and distributed programs. IPS provides performance data about the execution of *parallel systems* and performance analysis techniques that help to locate the program's bottlenecks. IPS is based on a hierarchical model of the program, which presents multiple levels of abstraction, providing multiple views of performance data. IPS consists of four components:

1. the instrumentation probes generating trace data;

2. the data pool storing the trace data;

3. the analyst which is a set of processes that summarizes and evaluates the measurement data; and

4. an interactive user interface for presenting the results.

IPS provides two automatic analysis techniques: Critical Path Analysis (CPA) and Phase Behavior Analysis. The CPA identifies the path through the program that consumed the most time. To perform the CPA, IPS constructs a program activity graph. The goal of the Phase Behavior Analysis is to identify phases in the program's execution history and focus on each phase as a separate problem. A phase is a period of time, in which a combination of performance metrics maintain consistent values. IPS can display a metrics table for a phase and display the portion of critical path that lies within the phase.

[86] B. P. Miller, C.-Q. Yang, IPS: *an interactive and automatic performance measurement tool for parallel and distributed programs*, in "Proceedings of the Seventh Conference on Distributed Memory Computer Systems", Vol. 7, 1987, pp. 482–489.

IPS is an interactive tool for performance measurement and analysis of parallel and distributed programs. IPS is based on two main principles. First, users should be supplied with the maximum information about the executing program. This information should be available from all levels of abstraction – from the statement level up to the process interaction level. To prevent the user from being inundated with irrelevant details, there must be a logical and intuitive organization to this data. Second, users should be supplied with answers, not numbers. The performance tool should be able to guide the user to the location of the performance problem, and describe the problem in terms of the source program.

[87] T. G. Moher, PROVIDE: *A Process Visualization and Debugging Environment*, IEEE Transactions on Software Engineering 14 (1988), pp. 849–857.

The first part of the article deals with the limitations of conventional debugging. The second part describes PROVIDE – a process visualization and debugging environment. The major features of the software tool are the concepts of "deferred binding program animation", which allows users to interactively change the depiction of program execution during the debugging task, and "process history consistency maintenance", which guarantees a consistent record of program execution in the face of changes to program instructions and run time data values.

[88] T. G. Moher, P. R. Wilson, *Offsetting Human Limits with Debugging Technology*, IEEE Software 8-3 (1991), pp. 11–13.

The debuggers described in this issue are visibility tools, which make hidden information available to programmers. The principles of visibility demonstrated by the article in this issue – multiple views at varying levels of abstraction, direct support for backward-chaining problem solving, isolation of concurrent threads of execution, and the like – will remain important as long as people are involved in the development phase.

[89] B. Mohr, *Standardization of Event Traces Considered Harmful (Is an Implementation of Object-Independent Event Trace Monitoring and Analysis Systems Possible?)*, in "Environments and Tools for Parallel Scientific Computing" (J. J. Dongarra, B. Tourancheau, Eds.), Elsevier Science Publishers, Amsterdam, 1993, pp. 103–124.

This article discusses approaches to implementing *object-independent* event trace monitoring and analysis systems. The term object-independent means that the system can be used for the analysis or arbitrary (non-sequential) computer systems, operating systems, programming languages and applications. Three main topics are addressed: (i) object-independent monitoring, (ii) standardization of event trace formats and access interfaces and (iii) the application-independent but problem oriented implementation of analysis and visualization tools.

Based on these approaches, the distributed hardware monitor system ZM4 and the
SIMPLE event trace analysis environment were implemented, and have been used in
many "real-world" applications throughout the last three years. An overview of the
projects in which the ZM4/SIMPLE tools were used is given in the last section.

[90] M. Moriconi, D. F. Hare, *Visualizing Program Designs Through* PE-
GASYS, IEEE Computer 18-8 (1985), pp. 72-85.

PEGASYS is a *programming environment* for the graphical analysis of systems. It
facilitates the explanation of program design. A program design is described by a
hierarchy of interrelated pictures. Each picture describes data and control dependencies among subprograms, processes, modules and data objects. PEGASYS checks
whether or not the pictures are syntactically meaningful, enforces design rules and
determines if the program meets its pictorial documentation.

[91] A. Müller, J. Winckler, S. Grzybek, M. Otte, F. Equoy, N. Higelin, *The
Program Animation System* PASTIS, The Journal of Visualization and
Computer Animation 2 (1991), pp. 26-33.

The software animation system PASTIS (Program Animation System with Interactive Solutions), designed for program animation with emphasis on the visualization
of the dynamic behavior of algorithms and data structures, is presented. Its main
properties are an unmodified source code of the visualized program, concurrent multiple views on algorithms and data structures, and interactive alterations of views
during run time. PASTIS is distinguished by high modularity and strict separation of
its components. This makes it particularly suitable for distributed computing environments. The interface between program and animation is a relational data model.
Animations are directed by a single tuple, or sets of tuples, called relations, or sets
of relations, called networks. Animations can be nested hierarchically.

[92] B. A. Myers, *Taxonomies of Visual Programming and Program Visualization*, Journal of Visual Languages and Computing 1 (1990), pp.
97-123.

The terms *Visual Programming* and *Program Visualization* are defined. The program
visualization systems reported in the literature are classified as systems using *static*
or *dynamic* displays for visualization of program *code, data*, or the *algorithm* of the
program.

[93] L. M. Ni, K. C. Tai, *Special Issue on Software Tools for Parallel Programming and Visualization: Guest Editor's Introduction*, Journal of
Parallel and Distributed Computing 9 (1990), pp. 101-102.

A report is given about current research and software tools, supporting the development of parallel programs. Ten papers were selected, and divided into four categories,
providing a brief summary of these papers categorized in the following topics:

- parallel languages and their implementation;
- design and simulation tools;
- static analysis tools; and
- monitoring and execution analysis tools.

[94] K. M. Nichols, *Performance Tools*, IEEE Software 7-3 (1990), pp. 21–30.

The following *performance tools* are described:

- NASS (Network Architecture Simulation System);
- Q+ (visual performance modeller) (Funka-Lea et al. [52]);
- PM (parallel-system profiler);
- PERSPECTIVE (performance, reliability evaluation);
- AXE (environment for concurrent systems);
- HYPERVIEW (performance analysis and visualization);
- SIMPLE/CARE (concurrent-application analyzer);
- POKER (parallel-programming environment) (Snyder [113]);
- TRIPLEX (parallel-execution monitor);
- ARTS (real-time scheduler analyzer/debugger); and
- JED (event display for multiprocessor systems) (Malony [78]).

For all their differences, the tools have some common features. Seven of these tools run under X Window. Another important characteristic shared by many of these tools is the view of the target as a system architecture, with interacting functional layers. Other common features are the use of graphs to depict underlying systems, the use of modelling hierarchy to ease reusability and comprehensibility, and an emphasis on identifying the bottlenecks of architectures, programming, or communications.

[95] K. M. Nichols, J. T. Edmark, *Modelling Multicomputer Systems with* PARET, IEEE Computer 21-5 (1988), pp. 39–48.

PARET (Parallel Architecture and Evaluation Tool) provides an interactive environment for the *design and study of multicomputer systems*. Multicomputers are non-shared memory multiprocessors of the MIMD class. For modelling parallel systems, PARET chooses simulation rather than detailed emulation. PARET depicts parallel programs as directed flow graphs. It uses them to model user programs, operating systems and hardware primitives. PARET uses a graphical user interface to maintain a consistent environment. PARET is intended to have the flexibility to serve as a multicomputer laboratory. It has a number of methods for measuring performance, including visual observation.

[96] K. Nichols, P. W. Oman, *Navigating Complexity to Achieve High Performance*, IEEE Software 8-5 (1991), pp. 12–15.

There are two major approaches to performance analysis: measuring and modelling. Measurement, embodied in hardware and software instrumentation and teaching, quantifies a real system. Modelling lets us simulate and analyze a real or proposed system. Two measurement techniques are described:

- Measuring and Analyzing Real Time Performance (Kenny, Lin [6]); and
- Finite-Element Analysis on a PC.

Additionally four modelling techniques are discussed:

- TRACEVIEW (Malony et al. [80]);
- PARAGRAPH (Heath, Etheridge [58]);
- Performability Modelling with ULTRASAN; and
- Interactive Visual Modelling from Performance Analysis (Funka-Lea et al. [52]).

[97] R. F. Paul, D. A. Poplawski, *Visualizing the Performance of Parallel Matrix Algorithms*, in "Proceedings of the Fifth Distributed Memory Computing Conference, Charleston, South Carolina, Vol. 2: Architectures, Software Tools and other General Issues" (Q. Stout, M. Wolfe, Eds.), IEEE Press, Washington DC, 1990, pp. 1207–1212.

The animation Tool MaTRIX (Matrix TRace In X) for performance evaluation of parallel algorithms for dense matrix operations is described. It portrays the execution of a program in the context of the application by displaying the primary matrix and showing which parts of the matrix are being operated on, which processors are operating on those parts, and what operations are being performed. Colors and patterns are being used to identify activity and differentiate between unique processors and various operations. The animation uses post-processed trace files generated during the execution of a program, thereby enabling the display to be run at various speeds. Coupled with displays of processor activity and utilization, the animation provides application-oriented performance information that is useful in determining causes of poor performance. The tool is written to use X Window and employ the tracing facilities in the PICL library, and is thereby portable to a wide range of parallel architectures and visual display devices.

[98] D. Pease, A. Ghafoor, I. Ahmad, D. L. Andrews, K. Foudil-Bey, T. E. Karpinski, M. A. Mikki, M. Zerrouki, PAWS: *A Performance Evaluation Tool for Parallel Computing Systems*, IEEE Computer 24–1 (1991), pp. 18–29.

PAWS (Parallel Assessment Window System) a *performance evaluation* tool for *parallel computing systems* is presented. PAWS consists of four tools:

1. The application characterisation tool translates applications written in a high-level language into a data dependence graph. This dataflow graph is a machine-independent intermediate representation.

2. The architecture characterisation tool allows the user to create descriptions of machines.

3. The interactive graphical display tool provides the user interface for accessing all PAWS tools.

4. The performance assessment tool allows the user to evaluate the performance of any application entered.

The performance metrics include speed-up curves (the average amount of computation performed in one step within unlimited processors), parallelism profile curves and execution profiles. It shows both the ideal parallelism inherent in the machine-independent dataflow graph, and the predicted parallelism of the dataflow graph on the target machine.

[99] W.-P. Po, T. Lewis, S. Thakkar, *System Performance Displayer: A performance Monitoring Tool*, Technical Report 91–30–6, Oregon State University Computer Science Department, 1991.

SPD (System Performance Displayer) is a tool for *interpretation* and *display* of performance data. It should help the user to detect bottlenecks. The target architecture is a shared memory multiprocessor system (from Sequent); the tool itself was implemented for Macintosh providing a user-friendly interface.

[100] S. Poinson, B. Tourancheau, X. Vigouroux, *Distributed Monitoring for Scalable Massively Parallel Machines*, in "Environments and Tools for Parallel Scientific Computing" (J. J. Dongarra, B. Tourancheau, Eds.), Elsevier Science Publishers, Amsterdam, 1993, pp. 85–101.

This paper is similar to the publication Tourancheau et al. [117], with more example views given.

[101] R. J. Pooley, *The Integrated Modelling Support Environment – A New Generation of Performance Modelling Tools*, in "Proceedings of the 5th International Conference on Modelling Techniques and Tools for Computer Performance Evaluation", Elsevier Science Publishers, Torino, Italy, 1991, pp. 1–15.

Performance engineering is the entire process of using performance techniques in hardware and software design thus including the modelling, the measurement and the use of the respective results. This paper presents the Integrated Modelling Support Environment (IMSE), which should support system modellers in the whole process of software engineering. IMSE contains a set of tools for modelling (process interaction tool PIT, QNET based on queuing network models and a Petri net tool by G. Chiola), workload analysis (WAT by M. Calzarossa and L. Massari) and system description (SP tool by P. H. Hughes). A generic graph manipulation facility (SDMF) provides integration and consistency in the interface of tools. An object management system ensures that common data structures have a consistent meaning to all tools. An *experimenter* (allowing the user to perform series of model solutions using one or more models), a *reporter* (generating reports on experiments) and an *animator* (using trace files produced by a simulator) are also included.

[102] D. A. Reed, R. D. Olson, R. A. Aydt, T. M. Madhyastha, T. Birkett, D. W. Jensen, B. A. A. Nazief, B. K. Totty, *Scalable Performance Environments for Parallel Systems*, in "Proceedings of the Sixth Distributed Memory Computing Conference, Portland, Oregon" (Q. Stout, M. Wolfe, Eds.), IEEE Press, Washington DC, 1991, pp. 562–569.

A prototype of a performance analysis environment is presented, in which performance data obtained from trace files is visualized, but also sonic data representation is supported.

[103] S. P. Reiss, PECAN: *Program Development Systems that Support Multiple Views*, IEEE Transactions on Software Engineering SE–11 (1985), pp. 276–285.

A programmer developing applications in the PECAN environment can observe his program via multiple views which are visual representations of abstract syntax trees. Currently two different views are available: a syntax directed editor and Nassi-Shneiderman flow graph representation. In addition some special semantic views are provided (symbol table view, data type view, expression view, flow view). Future implementations will include a data-flow view, a module-level abstraction view, and an execution view showing changes in the program during execution.

[104] P. N. Robillard, F. Coallier, D. Coupal, *Profiling Software Trough the Use of Metrics*, Software Practice and Experience 21 (1991), pp. 507–518.

An approach to software assessment using a new software tool that integrates most of the known static metrics is presented. This metrics are computed from the static analysis of the source code. Either all of or a subset of the metrics are selected to perform the assessment. Metric distributions are studied to determine the usual ranges of values. Out-of-range functions are identified. Percentile profiles give a project overview of the percentage of out of range functions. Quality factors are defined by grouping metrics. A normality profile gives the percentage of success for every quality factor. Finally, function couplings are evaluated from a 3D call graph representation.

[105] G.-C. Roman, K. C. Cox, *A Declarative Approach to Visualizing Concurrent Computations*, IEEE Computer 22-10 (1989), pp. 25–36.

A declarative approach for *visualizing concurrent computations* is described. The advantage of a declarative approach is that it does not involve modification of the program code, like the imperative approach does. The declarative approach treats the visualization of computations as the application of a function to the computational state yielding an image. This function in called the visualization function (v : *states* → *images*). The visualization function can be divided in two parts, the abstraction function (a : *states* → *objects*) and a rendering function (r : *objects* → *images*). The presented visualization methodology is based on program correctness. The same properties used to prove a program's correctness can be used to indicate which aspects should be represented in visualization. Further the structure of a property provides a guide to how that property should be visualized. Invariants would be visualized as stable patterns and progress properties would be visualized as evolving patterns.

[106] D. T. Rover, *Visualization of program performance of concurrent computers*, Ph.D. Thesis, Iowa State University, Ames IA, 1989.

[107] D. T. Rover, M. B. Carter, J. L. Gustafson, *Performance Visualization of* SLALOM, in "Proceedings of the Sixth Distributed Memory Computing Conference, Portland, Oregon" (Q. Stout, M. Wolfe, Eds.), IEEE Press, 1991, pp. 543–550.

SLALOM (Scalable Language-independent Ames laboratory One-minute Measurement) is a benchmark program for parallel architectures (including SIMD and MIMD paradigms). This article describes how visualization of SLALOM execution reveals the complex behavior in parallel systems (in particular the nCUBE 2 and the MasPar MP-1). PARAGRAPH (Heath, Etheridge [57]) and VISTA (Tuchman et al. [118]) are used for visualization.

[108] D. T. Rover, G. M. Prabhu, T. C. Wright, *Characterizing the performance of concurrent computers: a picture is worth a thousands numbers*, in "Proceedings of the Fourth Conference on Hypercubes, Concurrent Computers, and Applications, Vol. 2" (J. Gustafson, Ed.), IEEE Press, Los Altos CA, 1989, pp. 245–248.

[109] D. T. Rover, T. C. Wright, *Pictures of Performance: Highlighting Program Activity in Time and Space*, in "Proceedings of the Fifth Distributed Memory Computing Conference, Charleston, South Carolina, Vol. 2: Architectures, Software Tools and other General Issues" (D. W. Walker, Q. F. Stout, Eds.), IEEE Press, Washington DC, 1990, pp. 1228–1233.

Program performance evaluation in concurrent computer systems is a difficult task that requires methods and tools for observing, analysing, and displaying system performance. This paper describes a "machine perspective" for interpreting performance information, and the pictorial representation that has been developed to support this perspective. Performance data obtained from invasive instrumentation is reduced and converted into a pictorial form that highlights important aspects of program states during execution. Images are used to visually display both temporal and spatial information describing system activity. Phenomena such as hot spots of activity are easily observed, and in some cases, patterns inherent in the application algorithms being studied are highly visible. The approach couples qualitative observations and quantitative measurements to create a coherent representation of performance. A prototype of the prescribed tool has been developed and used to examine program performance. A case study illustrating the utility of the approach is presented.

[110] R. V. Rubin, L. Rudolph, D. Zernik, *Debugging Parallel Programs in Parallel*, ACM SIGPLAN Notices 24–1 (1989), pp. 216–225.

A Monitoring Animating Debugging (MAD) System, the Makbilan machine, for *parallel programs* is described. The Makbilan machine is a parallel monitor machine to "spy" on the main machine. It is responsible for filtering, for updating the shared database of information, for compound event analysis, for invoking the view handler and for performing graphical operations. A parallel graphical system is necessary to support the display of parallel program execution. The MAD system supports the task-state view, a dynamic process tree view, processing element's activity view, the work tree view and the processing element's task view.

[111] L. Rudolph, Z. Segall, PIE: *A Programming and Instrumentation Environment for Parallel Processing*, IEEE Software 2–6 (1985), pp. 22–37.

PIE is a parallel programming environment supporting the programmer in designing performance efficient parallel applications on shared memory architectures. Debugging and the detection (or even better the avoidance) of system bottlenecks are the most important tasks.

[112] T. Shimomura, S. Isoda, *Linked-List Visualization for Debugging*, IEEE Software 8–3 (1991), pp. 44–51.

The version 2 of VIPS a debugging tool for linked lists is described. The disadvantage from version 1 (Isoda et al. [65]) was that it had only one display mode, which showed each node's data values. Therefore only parts of a large linked list could be displayed. Version 2 can selectively display a linked list. VIPS also uses multiple windows to display various aspects of execution simultaneously. The rest of the article describes the usage of the tool and how to find errors.

[113] L. Snyder, *Parallel programming and the* POKER *programming environment*, IEEE Computer 17–7 (1984), pp. 27–37.

The non-shared memory model of parallelism is addressed in the development of POKER, a parallel programming environment providing an interactive graphics system for defining the communication graph of the algorithm. The layout of the graph is pre-given as a two-dimensional lattice in which the communication patterns can be embedded. Sequential code is programmed in the usual way, and those code blocks can be assigned to processors (by simply entering the name of the procedure in the box representing the processing element in the communication graph). After making some declarations and translations the resulting program is executed and traced variables can be viewed, the execution can be stopped, variable values can be changed and the execution will continue using the new values (the new values are "poked" back into the processor memories).

[114] D. Socha, M. L. Bailey, N. Notkin, VOYEUR: *Graphical Views of Parallel Programs*, ACM SIGPLAN Notices 24–1 (1989), pp. 206–215.

VOYEUR is a prototype system designed to construct application-specific *visual views of parallel programs*. VOYEUR is used in *monitoring* and *debugging* parallel systems. VOYEUR uses a class hierarchy of views. The vector and icon views are members of a class of x, y views. These views represent a state by placing graphical elements on an xy coordinate space. The icon view uses a grid. The vector view uses real valued coordinates. The trace view and link-list view are members of the text class views. Views of this class use lines to link boxes containing textual fields.

[115] J. T. Stasko, TANGO: *A Framework and System for Algorithm Animation*, IEEE Computer 23–9 (1990), pp. 27–39.

The author introduces a system for *algorithm animation* called TANGO (Transition based ANimation GeneratiOn). Algorithm animation is the process of abstracting a program's data, operations and semantics and creating dynamic graphical views of those abstractions. Algorithm animation encompasses program animation and datastructure rendering. Stasko developed an algorithm animation framework that helps to define abstract operations or events in a program, to design animation actions to simulate those operations and to map the abstract operations to their corresponding animation actions. Stasko developed the path-transition paradigm for animation design. It provides a consistent way to define graduate changes or transitions in an animated view. The paradigm helps the designer to interpolate between animation states.

[116] J. M. Stone, *A Graphical Representation of Concurrent Processes*, ACM SIGPLAN Notices 24–1 (1989), pp. 226–235.

A graphical representation called a concurrency map is presented. The concurrency map expresses potential concurrency, and it is both a data structure for controlling replay and a graphical method of *representing concurrent processes*. The map displays the process histories as event streams on a time grid. Each event stream is divided into dependence blocks, which are defined by the interprocess dependencies.

Time dependence is expressed by an arrow from the end of a block in one process to the beginning of a block in another process. The collection of transformations of a map shows all the event orderings that are consistent with the given time dependencies. The interpretation of the map is the set of event orderings that preserve the process histories and the interprocess time dependencies.

[117] B. Tourancheau, X. Vigouroux, M. G. van Riek, *The Massively Parallel Monitoring System – a truly parallel approach to parallel monitoring*, in "Performance Measurement and Visualization of Parallel Systems" (G. Haring, G. Kotsis, Eds.), Elsevier Science Publishers, Amsterdam, 1993, pp. 1–17.

The monitoring of parallel systems involves the collection, interpretation and display of information concerning the execution of parallel programs. When advancing from parallel to massively parallel systems, the monitoring tools have to face the problem of scalability.

Concerning the user interface, the "displaying" of information (graphics, sound, etc.) must follow a more hierarchical and abstract point of view and on the user response-time point of view, the tool must implement a scalable no-bottleneck software architecture. In the paper, the authors try to answer that problem with a distributed monitoring solution, where most of the data treatment is done on the parallel target machine and where the classical data gathering bottleneck is avoided with distributed storage and a parallel treatment of the interpretation. The *views* are also handled by different processors on different external displays. This approach thus becomes fully scalable (in terms of collecting power, storage, computing and displaying power). The only assumption is, that the massively parallel target machine is able, for at least a part of its nodes, to run UNIX-like system calls. It is realistic regarding what will be provided in the next generation of parallel machines. The authors also present the protocol established for the data communications in an implementation which runs on a LAN of SUN workstations using PVM, the Parallel Virtual Machine system.

[118] A. Tuchman, D. Jablonowski, G. Cybenko, *A System for Remote Data Visualization*, CSRD Report 1067, Center for Supercomputing Research and Development, University of Illinois at Urbana-Champaign, 1991.

VISTA is an environment for remote data visualization. Program data is shown during execution (*simulation time animation*), but the executing application might be replaced by a trace file. The user interactively chooses the data and the type of display. The system was designed to be modular, providing also tailored modules for different environments, machines and applications.

[119] E. R. Tufte, *The Visual Display of Quantitative Information*, Graphics Press, Cheshire, 1983.

The first part of the book reviews the graphical practice of the two centuries since Playfair[4]. Both the graphical glories and the lost opportunities are presented. The second part of the book provides a language for discussing graphics and a practical

---

[4]William Playfair (1759–1823) developed or improved upon nearly all the fundamental graphical designs, seeking to replace conventional tables of numbers with the systematic visual representations of his "linear arithmetic"

theory of data graphics. Applying to most visual displays of quantitative information, the theory leads to changes and improvements in design, suggests why some graphics might be better than others, and generates new types of graphics. The emphasis is on maximizing principles, empirical measures of graphical performance, and the sequential improvement of graphics through revisions and editing. Insights into graphical design are to be gained from theories of what is excellent in art, architecture, and prose.

[120] S. J. Turner, W. Cai, *The "Logical Clock" Approach to the Visualization of Parallel Programs*, in "Performance Measurement and Visualization of Parallel Systems" (G. Haring, G. Kotsis, Eds.), Elsevier Science Publishers, Amsterdam, 1993, pp. 45–66.

A fundamental problem in the visualization of parallel program behavior is that of minimizing the intrusive nature of the monitoring tool (the so-called *probe effect*). This paper describes an approach to the monitoring of parallel programs, which aims to minimize the amount of intrusion by introducing a *logical clock* for each process, and controlling inter-process communication according to logical time rather than real time. Since the logical time of a process now plays the same role as that of real time when running without the monitor, a high degree of transparency is achieved.

The paper then discusses how the logical clocks mechanism is being incorporated into a set of graphical tools for the development and visualization of parallel programs. Different levels of abstraction for visual programming are provided, with the central level based on the idea of *concurrency maps*. The same diagrams that are constructed during the development of a program are also used to display feedback information from the monitor, using the correspondence that exists between concurrency maps and the logical clocks approach. In this way, the graphical environment is able to provide facilities for the interactive debugging and performance analysis of parallel programs.

[121] E. M. Williams, G. B. Lamont, *A Real-Time Parallel Algorithm Animation System*, in "Proceedings of the Sixth Distributed Memory Computing Conference, Portland, Oregon" (Q. Stout, M. Wolfe, Eds.), IEEE Press, Washington DC, 1991, pp. 551–561.

Algorithm animation is a visualization method used to enhance understanding of the functioning of an algorithm or program. Visualization is used for many purposes, including education, algorithm research, performance analysis, and program debugging. This research applies algorithm animation techniques to programs developed for parallel architectures, with specific emphasis on the Intel iPSC/2 hypercube. Current investigations focus in two different areas: performance data display and animations of specific algorithms or classes of algorithms. This research builds on these efforts to provide a system that is able to both display performance data from parallel programs and support the creation of animations for specific algorithms. There are three goals for this visualization system:

- Data should be displayed as it is generated.
- The interface to the target program should be transparent, allowing the animation of existing programs.
- The system must be flexible enough to animate any algorithm.

The resulting system incorporates, integrates and extends two systems: the AFIT Algorithm Animation Research Facility (AAARF) and the Parallel Resource Analysis Software Environment (PRASE). Since performance data is an essential part of analysing any parallel program, multiple views of the performance data are provided as an elementary part of the system. In addition to the animation system, a method for developing the animations is discussed. This method is applicable to animate any type of program, sequential or parallel.

[122] L. D. Wittie, *Debugging Distributed C Programs by Real Time Replay*, ACM SIGPLAN Notices 24-1 (1989), pp. 57–67.

BUGNET is a tool for *debugging* C programs distributed within a networked UNIX system. It monitors the execution of programs collecting information about inter-process communication, I/O events and execution traces. The user can obtain this information via a graphical interface.

[123] E. Zabala, R. Taylor, *Process and Processor Interaction: Architecture Independent Visualization Schema*, in "Environments and Tools for Parallel Scientific Computing" (J. J. Dongarra, B. Tourancheau, Eds.), Elsevier Science Publishers, Amsterdam, 1993, pp. 55–71.

This paper proposes a model that combines visual and audio stimuli to organize and present trace data available to the user of parallel systems. The psychological basis for this model is presented, and the implications for applicability and implementation explored.

A practical implementation of this model *Maritxu* is introduced. Maritxu focusses on the presentation of the run time behavior of networks of parallel processors. It provides the user with total control over the visualization process, and as such is an ideal tool for both analysis and experiment. The paper concludes by examining the effectiveness of this approach when used to analyse real data. Recommendations for further development of the model in relation to this implementation are presented and discussed.

Visual Programming

[124] G. P. Brown, R. T. Carling, C. F. Herot, D. A. Kramlich, P. Souza, *Program Visualization: Graphical Support for Software Development*, IEEE Computer 18-8 (1985), pp. 27–35.

Performance visualization should accompany the whole software life cycle. In order to fulfil this claim several categories of program illustration must be supported (such as system requirements diagrams, program function diagrams, program structure diagrams, communication diagrams, composed and typeset program text, program comments and commentaries, diagrams of flow of control, of structured data, of persistent data, and of the program in the host environment). Within the PV project a visualization system should be developed offering the displays mentioned above. Most emphasis is given to dynamic data structure displays that allow the observation of changes during program execution.

[125] S. S. Yau, X. Jia, *Visual Languages and Software Specifications*, in "Proceedings of the International Conference on Computer Languages"

(P. Hsia, D. C. Rine, C. V. Ramamoorthy, Eds.), IEEE Press, Miami Beach, 1988, pp. 322–327.

Due to the rapid development of VLSI, microprocessor, and computer graphics technologies in recent years, high resolution, high speed graphics-based workstations now become economically feasible. These workstations have made so-called visual languages practical. In this paper the important issues and major features of visual languages are summarized and their impact on software engineering, especially software specification techniques, are discussed. The potential of using visual languages for software specifications to improve software reliability, modifiability, reusability and understandability is also considered.

Solution of ODEs

[126] O. Abou-Rabia, *A class of variable stepsize formulas for the parallel solution of ODEs*, Mathematics and Computers in Simulation 31 (1989), pp. 165–169.

[127] C. Addison, J. Allwright, N. Binsted, N. Bishop, B. Carpenter, P. Dalloz, D. Gee, V. Getov, T. Hey, R. Hockney, M. Lemke, J. Merlin, M. Pinches, C. Scott, I. Wolton, *The Genesis Distributed Memory Benchmarks. Part 1: Methodology and General Relativity benchmark with results for the* SUPRENUM *Computer*, Concurrency, Practice and Experience 5 (1993), pp. 1–22.

[128] R. Augustyn, M. Karg, C. W. Ueberhuber, *Parallel Solution of ODE IVPs – A Literature Survey*, ACPC/TR 92-21, Austrian Center for Parallel Computation, Vienna, 1992.

[129] R. Augustyn, A. R. Krommer, C. W. Ueberhuber, *Effizient-Portable Programmierung von Parallelrechnern*, Report 85/91, Institute for Applied and Numerical Mathematics, Technical University Vienna, 1991.

[130] R. Augustyn, C. W. Ueberhuber, *Parallel Defect Correction Algorithms for Ordinary Differential Equations*, Technical Report ACPC/TR 92-22, Austrian Center for Parallel Computation, Vienna, 1992.

[131] R. Augustyn, C. W. Ueberhuber, *Simulationssoftware zur Untersuchung paralleler IDeC-Varianten*, Report 89/92, Institute for Applied and Numerical Mathematics, Technical University Vienna, 1992.

[132] R. Augustyn, C. W. Ueberhuber, *Influence of Granularity on Parallel IDeC Methods*, Technical Report 108/93, Institute for Applied and Numerical Mathematics, Technical University Vienna, 1993.

[133] R. Augustyn, C. W. Ueberhuber, *Performance Modelling and Evaluation of Parallel IDeC Methods*, Technical Report SciPaC/TR 93-7, Scientific Parallel Computation Group, Technical University Vienna, 1993.

[134] R. Augustyn, C. W. Ueberhuber, *Parallel Algorithms and Software for ODE IVPs*, Report 88/92, Institute for Applied and Numerical Mathematics, Technical University Vienna, 1992.

[135] R. Augustyn, C. W. Ueberhuber, *Load Distribution in Parallel IDeC Methods*, Technical Report SciPaC/TR 93-4, Scientific Parallel Computation Group, Technical University Vienna, 1993.

[136] W. Auzinger, R. Frank, G. Kirlinger, *Asymptotic Error Expansions for Stiff Equations: Applications*, Computing 43 (1990).

[137] O. Axelsson, B. Polman, *On approximate factorization methods for block matrices suitable for vector and parallel processors*, Lin. Alg. and its Applic. 77 (1986), pp. 3–26.

[138] D. P. Bertsekas, J. N. Tsitsiklis, *Parallel and Distributed Computation – Numerical Methods*, Prentice-Hall, Englewood Cliffs, 1989.

[139] L. G. Birta, O. Abou-Rabia, *Parallel block predictor-corrector methods for ODEs*, IEEE Transactions on Computers 36 (1987), pp. 299–311.

[140] K. Burrage, *A special family of Runge-Kutta methods for solving stiff differential equations*, BIT 18 (1978), pp. 22–41.

[141] K. Burrage, *The error behaviour of a general class of predictor-corrector methods*, Applied Numerical Mathematics 8 (1991), pp. 201–216.

[142] J. C. Butcher, *Towards efficient implementation of singly-implicit methods*, ACM Trans. Math. Softw. 14 (1988), pp. 68–75.

[143] J. R. Cash, *Block Runge-Kutta methods for the numerical integration of initial value problems in ordinary differential equations*, Mathematics of Computation 40 (1983), pp. 175–191.

[144] J. R. Cash, *Block embedded explicit Runge-Kutta methods*, Computers and Mathematics with Applications 11 (1985), pp. 395–409.

[145] M. T. Chu, H. Hamilton, *Parallel solution of ODEs by multi-block methods*, SIAM Journal on Scientific and Statistical Computing 8 (1987), pp. 342–353.

[146] J. J. Dongarra, D. C. Sorenson, *Linear algebra on high performance computers*, Appl. Math. and Comp. 20 (1986), pp. 57–88.

[147] K. Fasching, *Implementierung von Defektkorrektur-Algorithmen für Systeme steifer Differentialgleichungen*, Diplomarbeit, Technical University Vienna, 1990.

[148] R. Frank, J. Hertling, H. Lehner, *B-convergence properties of defect correction methods*, Numer. Math. 49 (1986), pp. 139–188.

[149] R. Frank, F. Macsek, C. W. Ueberhuber, *Asymptotic error expansions for defect correction iterates*, Computing 32 (1984), pp. 115–125.

[150] R. Frank, F. Macsek, C. W. Ueberhuber, *Iterated Defect Correction for Differential Equations, Part II: Numerical Experiments*, Computing 33 (1984), pp. 107–129.

[151] R. Frank, J. Schneid, C. W. Ueberhuber, *The concept of B-convergence*, SIAM J. Numer. Anal. 18 (1981), pp. 753–780.

[152] R. Frank, J. Schneid, C. W. Ueberhuber, *Order results for implicit Runge-Kutta methods applied to stiff systems*, SIAM J. Numer. Anal. 22 (1985), pp. 515–534.

[153] R. Frank, J. Schneid, C. W. Ueberhuber, *Stability properties of implicit Runge-Kutta methods*, SIAM J. Numer. Anal. 22 (1985), pp. 497–515.

[154] R. Frank, J. Schneid, C. W. Ueberhuber, *B-convergence: a survey*, Applied Numerical Mathematics 5 (1989), pp. 51–61.

[155] R. Frank, J. Schneid, C. W. Ueberhuber, *Quantitative Analyse von Verfahren für steife Differentialgleichungen*, Z. Angew. Math. Mech. 69 (1989), pp. 135–136.

[156] R. Frank, C. W. Ueberhuber, *Iterated defect correction for the efficient solution of stiff systems of ordinary differential equations*, BIT 17 (1977), pp. 146–159.

[157] R. Frank, C. W. Ueberhuber, *Collocation and iterated defect correction*, in "Numerical treatment of differential equations" (R. Bulirsch, R. D. Grigorieff, J. Schroeder, Eds.), Lecture Notes in Mathematics 631, Springer-Verlag, Berlin Heidelberg New York Tokyo, 1978, pp. 19–34.

[158] R. Frank, C. W. Ueberhuber, *Iterated Defect Correction for Differential Equations, Part I: Theoretical Results*, Computing 20 (1978), pp. 207–228.

[159] C. W. Gear, *The Potential for Parallelism in Ordinary Differential Equations*, Technical Report UIUCDCS-R-86-1246, Univ. Illinois at Urbana-Champaign, 1986.

[160] C. W. Gear, *Massive parallelism across the method in ODEs*, Technical Report UIUCDCS-R-88-1442, Univ. Illinois at Urbana-Champaign, 1988.

[161] C. W. Gear, *Parallel methods for ordinary differential equations*, Calcolo 25 (1988), pp. 1–20.

[162] C. W. Gear, F.-L. Juang, *The speed of waveform methods for ODEs*, in "Applied and Industrial Mathematics" (R. Spigler, Ed.), Kluwer Academic Pub., Netherlands, 1991, pp. 37–48.

[163] C. W. Gear, D. R. Wells, *Multirate linear multistep methods*, BIT 24 (1984), pp. 484–502.

[164] R. W. Hockney, C. R. Jesshope, *Parallel Computers 2*, Adam Hilger, Bristol Philadelphia, 1988.

[165] F.-L. Juang, *Accuracy increase in waveform relaxation*, Technical Report UIUCDCS-R88-1466, Univ. Illinois at Urbana-Champaign, 1988.

[166] F.-L. Juang, C. W. Gear, *Accuracy increase in waveform Gauss Seidel*, Technical Report UIUCDCS-R89-1518, Univ. Illinois at Urbana-Champaign, 1989.

[167] M. Karg, R. Augustyn, C. W. Ueberhuber, *IDeC-Verfahren mit parallelen Defektkorrekturebenen auf Rechnern mit globalem Speicher*, Bericht Nr. 98/92, Institut für Angewandte und Numerische Mathematik, 1992.

[168] M. Karg, R. Augustyn, C. W. Ueberhuber, *IDeC-Verfahren mit parallelen Zeitschritten auf Rechnern mit globalem Speicher*, Bericht Nr. 95/92, Institut für Angewandte und Numerische Mathematik, 1992.

[169] M. Karg, R. Augustyn, C. W. Ueberhuber, *IDeC-Verfahren mit parallelen Defektkorrekturebenen auf Rechnern mit verteiltem Speicher*, Bericht Nr. 100/93, Institut für Angewandte und Numerische Mathematik, 1993.

[170] M. Karg, R. Augustyn, C. W. Ueberhuber, *IDeC-Verfahren mit parallelen Zeitschritten auf Rechnern mit verteiltem Speicher*, Bericht Nr. 99/93, Institut für Angewandte und Numerische Mathematik, 1993.

[171] A. R. Krommer, C. W. Ueberhuber, *PF 90 – A Description Language for Parallel Algorithms*, Technical Report ACPC/TR 91-12, Austrian Center for Parallel Computation, Vienna, 1991.

[172] A. R. Krommer, C. W. Ueberhuber, *Dynamic Load Balancing: An Overview*, Technical Report ACPC/TR 92-18, Austrian Center for Parallel Computation, Vienna, 1992.

[173] F. Le Gall, G. Mouney, *About the generation of parallel block predictor-corrector algorithms for ODEs*, International Journal of Computer Mathematics 36 (1990), pp. 209–219.

[174] E. Lelarasmee, A. E. Ruehli, A. L. Sangiovanni-Vincentelli, *The waveform relaxation method for time-domain analysis of large scale integrated circuits*, IEEE Trans. CAD Integr. Circuits Systems 1 (1982), pp. 131–145.

[175] W. L. Miranker, *A survey of paralellism in numerical analysis*, SIAM Review 13 (1971), pp. 524–547.

[176] J. I. Montijano, *Estudio de los methodos SIRK para la resolucion numerica de ecuaciones diferenciales de tipo stiff*, Thesis, University of Zaragoza, 1983.

[177] J. Nievergelt, *Parallel methods for integrating ordinary differential equations*, Communications of the ACM 7 (1964), pp. 731–733.

[178] J. M. Ortega, R. G. Voigt, *Solution of partial differential equations on vector and parallel computers*, SIAM Review 27 (1985), pp. 149–240.

[179] J. Sand, S. Skelboe, *Stability of backward Euler multirate methods and convergence of waveform relaxation*, BIT 32 (1992), pp. 350–366.

[180] L. F. Shampine, H. A. Watts, *Block implicit one-step methods*, Mathematics of Computation 23 (1969), pp. 731–740.

[181] L. F. Shampine, H. A. Watts, *Global Error Estimation for Ordinary Differential Equations*, ACM Transactions on Mathematical Software 2 (1976), pp. 172–186.

[182] R. D. Skeel, *Waveform iteration and the shifted Picard splitting*, SIAM Journal on Scientific and Statistical Computing 10 (1989), pp. 756–776.

[183] S. Skelboe, *Stability properties of backward differentiation multirate formulas*, Applied Numerical Mathematics 5 (1989), pp. 151–160.

[184] S. Skelboe, P. U. Andersen, *Stability properties of backward Euler multirate formulas*, SIAM J. Sci. Stat. Comput. 10 (1989), pp. 1000–1009.

[185] B. P. Sommeijer, W. Couzy, P. J. van der Houwen, *A-stable parallel block methods for ordinary and integro-differential equations*, Applied Numerical Mathematics 9 (1992), pp. 267–281.

[186] H. W. Tam, *One-stage parallel methods for the numerical solution of ordinary differential equations*, SIAM Journal on Scientific Statistical Computing 13 (1992), pp. 1039–1061.

[187] H. W. Tam, *Two-stage parallel methods for the numerical solution of ordinary differential equations*, SIAM Journal Scientific Statistical Computing 13 (1992), pp. 1062–1084.

[188] C. W. Ueberhuber, *Implementation of Defect Correction Methods for Stiff Differential Equations*, Computing 23 (1979), pp. 205–232.

[189] P. J. van der Houwen, B. P. Sommeijer, *Block Runge-Kutta methods on parallel computers*, Zeitschrift für angewandte Mathematik und Mechanik 72 (1992), pp. 3–18.

[190] D. R. Wells, *Multirate linear multistep methods for the solution of systems of ordinary differential equations*, Technical Report UIUCDCS-R-82-1092, Univ. Illinois at Urbana-Champaign, 1982.

[191] J. White, A. L. Sangiovanni-Vincentelli, F. Odeh, A. Ruehli, *Waveform relaxation: theory and practice*, Transactions of the Society for Computer Simulation 2 (1985), pp. 95–133.

[192] P. B. Worland, *Parallel methods for the numerical solution of ordinary differential equations*, IEEE Transactions on Computers 25 (1976), pp. 1045–1048.

[193] H. Zima, B. Chapman, *Supercompilers for Parallel and Vector Computers*, ACM Press, New York, 1991.

Numerical Integration

[194] A. R. Krommer, C. W. Ueberhuber, *Parallel Numerical Integration*, Technical Report SciPaC/TR 93-1, Scientific Parallel Computation Group, Technical University Vienna, 1993.

[195] A. R. Krommer, C. W. Ueberhuber, *A Simulator System for AAA Software Development*, Report 104/93, Institute for Applied and Numerical Mathematics, Technical University Vienna, 1993.

[196] R. Piessens, E. de Doncker-Kapenga, C. W. Ueberhuber, D. H. Kahaner, QUADPACK – *A Subroutine Package for Automatic Integration*, Springer-Verlag, Berlin Heidelberg New York Tokyo, 1983.

Linear Algebra

[197] J. J. Dongarra, J. DuCroz, I. Duff, S. Hammarling, *A Set of Level 3 Basic Linear Algebra Subprograms*, ACM Trans. Math. Software 16 (1990), pp. 1–17.

[198] J. J. Dongarra, J. DuCroz, S. Hammarling, R. Hanson, *An Extended Set of Fortran Basic Linear Algebra Subprograms*, ACM Trans. Math. Software 14 (1988), pp. 1–32.

[199] C. Lawson, R. Hanson, D. Kincaid, F. Krogh, *Basic Linear Algebra Subprograms for Fortran Usage*, ACM Trans. Math. Software 5 (1979), pp. 308–329.

Index

Lecture Notes in Computer Science

For information about Vols. 1–693
please contact your bookseller or Springer-Verlag